Investigating Mechanical Failures

Investigating Mechanical Failures

THE METALLURGIST'S APPROACH

Bob Ross

Managing Director of Ross Materials Technology, Glasgow, UK

CHAPMAN & HALL

London · Glasgow · Weinheim · New York · Tokyo · Melbourne · Madras

Published by Chapman & Hall, 2–6 Boundary Row, London SE1 8HN, UK

Chapman & Hall, 2–6 Boundary Row, London SE1 8HN, UK

Blackie Academic & Professional, Wester Cleddens Road, Bishopbriggs, Glasgow G64 2NZ, UK

Chapman & Hall GmbH, Pappelallee 3, 69469 Weinheim, Germany

Chapman & Hall USA, One Penn Plaza, 41st Floor, New York NY 10119, USA

Chapman & Hall Japan, ITP-Japan, Kyowa Building, 3F, 2-2-1 Hirakawacho, Chiyoda-ku, Tokyo 102, Japan

Chapman & Hall Australia, Thomas Nelson Australia, 102 Dodds Street, South Melbourne, Victoria 3205, Australia

Chapman & Hall India, R. Seshadri, 32 Second Main Road, CIT East, Madras 600 035, India

First edition 1995

Typeset in 10/12 Times by Mews Photosetting, Beckenham, Kent
Printed in Great Britain at the University Press, Cambridge.

ISBN 0 412 54920 4

A catalogue record for this book is available from the British Library

Library of Congress Catalog Card Number: 94-68721

Printed on permanent acid-free text paper, manufactured in accordance with the proposed ANSI/NISO Z39.48-1992 and ANSI/NISO Z39.48-1984 (Permanence of Paper).

Contents

Preface

Trained by Rolls Royce aeronautical engineers in Scotland first as an industrial chemist, then as a metallurgist, I was fully involved in failure investigation and the solving of material and production problems. This training took place under the auspices of the late legendary John McEwan, who taught me first to 'see', then to 'evaluate' and finally to write concise reports with a meaningful conclusion.

My academic training was gained at the Royal Technical College of Glasgow, better known as the 'Glasgow Tech', now the University of Strathclyde, and consisted of night classes, three nights a week after a day's work. Approximately 50% of lecturers were part-time in those days, working in management-type jobs during the day, which ensured considerable practical knowledge was imparted along with the theory in the subject of practical metallurgy.

It is my belief that the present education system makes it more difficult for students to gain this combination of practical and theoretical experience. Courses today tend to be either day-release or more usually full-time, taught by lecturers who are also full-time with limited outside experience. Consequently students graduate knowing very little of the practical, down-to-earth metallurgy which is essential when solving 'real' problems. It is my opinion that the consequences of this are much more serious than is normally realized, resulting in very high costs to industry with attendant health and safety implications. One of the reasons for writing this book is to attempt to redress this balance.

The Rolls Royce experience included chemistry and the control of electroplating, chemical analysis of materials associated with engineering and metallurgical investigation or mechanical failures in the field and on the shop floor. As a metallurgist located within the Planning Department to advise or control all operations pertaining to materials, the responsibility involved developing new types of production equipment. It also taught me to communicate with nonmetallurgical personnel, and highlighted the need to translate metallurgical vernacular into plain English.

In 1968 I set up my own consultancy with metallurgical and mechanical testing and chemical analysis facilities. This now has a world-wide reputation for failure investigation and materials problem-solving, with particular reference to oil rig anchor chain investigations. I have visited all five continents as far west as Honolulu, and as far east as Tokyo. Over the years I have gained a reputation for 'seeing' metallurgical problems rapidly, and producing reports which are understood and always contain a firm conclusion.

I am the author of *Metallic Materials Specification Handbook*, the first edition being published in 1968, with the fourth in 1992. This lists and classifies metal specifications, trade names, etc. The second book, *Handbook of Metal Treatments and Testing*, is now in its second edition, and lists and briefly describes over 600 techniques used by those involved in all aspects of metals.

This new book was compiled over a two-year period, partly from an original transcript of recordings made in the Scottish hills on a hand-held recorder. Ideas had been collected and discussed with many friends, colleagues, clients and critics. It could not have been produced without their help, comments and criticism. Those involved are too many to be listed, but it is hoped that the result will be useful to metallurgical students, engineers and others involved in mechanical failures.

Regarding the production of artwork, I wish to record my thanks to Joe Donachie of Tighnabruaich, Argyll, Scotland.

PART ONE

Preparation, Procedure and Presentation

Mental and physical preparation

1

This is a book on mechanical failures, their detection and investigation and the capture of over 30 years of practical experience in the form of case histories.

Dictionary definitions of 'failure' range from 'unsuccessful performance of functions' through 'non-performance', 'breakdown' and 'ill-success' to 'impairment of function' and can relate to any form of mechanical, electrical, medical, legal and/or chemical, situation.

This book confines itself to the solution of mechanical-type problems which cannot be solved by the use of engineering ability alone, but where help is required from the metallurgist.

THE DECLINE OF PRACTICAL TRAINING

Over the last decade it has become apparent that the cost and quality of some metallurgical advice and failure reporting leaves much to be desired. This is not unconnected with the fact that there are now quite limited opportunities for the recently qualified metallurgist to work alongside a practical qualified metallurgist. Many college lecturers are second or third generation academics, with little or no field experience.

It is not really possible to self-teach practical metallurgy. Thus there are now limited opportunities compared to those available to the author, who had guidance from several highly practical, qualified, experienced metallurgical gurus as advisers on a day-to-day basis. The fact that there are relatively few establishments with metallurgical laboratories does not mean that there is little or no need for good metallurgical advice. Welding, heat treatment, electroplating, painting and finishing together with the choice and application of materials all require a metallurgical input at some stage.

The limitations of present-day education in metallurgy was a major factor in producing this book. It is hoped that the metallurgical student can make use of the information recorded to relate the necessary academic information to the practical aspects of everyday metallurgy. It is offered as a resource to stand alongside the more academic textbooks.

An attempt has been made to keep the technical content at a level which can be understood by engineers from a wide range of disciplines, thus, hopefully, allowing pertinent questions to be asked during discussion of mechanical failures with the qualified metallurgist.

MENTAL ATTITUDES TOWARDS FAILURE INVESTIGATION

In my experience there are a limited number of reasons why a problem should be investigated, and the client with the failure should perhaps give some consideration to the motivation behind an investigation before setting out on what can be an expensive exercise. Such lack of objectivity has led to many metallurgical investigations being carried out and reported on where one or more relevant features had been overlooked.

The following are the most common reasons for requesting a technical enquiry, not given in any order of priority, as this will vary depending on the circumstances and very often the personalities involved.

- **For scientific purposes**. This is a valid reason where, for example, a large company has a problem which has not necessarily resulted in injury or financial loss. The company however, might decide as a matter of scientific curiosity to have an investigation carried out with the information recorded or circulated.
- **In order to apportion blame**. Where there has been financial loss, injury or damage investigation may be justified for no other reason than to obtain costs. This can lead to pressure on the investigator to examine only those aspects which would apportion blame in the desired direction, and to ignore aspects which do not suit the client.
- **To identify and eliminate the cause of failure**. This is to ensure that procedures are applied so that no similar failure can occur. This might be in an area of operational procedure, change of material, design, etc. Of all the reasons for carrying out an investigation, this is probably the most important, particularly where injury or death has resulted or the cost has been high. It also highlights the necessity for the investigator to always at least consider making recommendations in the report in addition to defining the cause of failure.
- **To improve performance**. This is an excellent reason for an investigation if the component or system can be made to last longer or to become more efficient, supply more power, etc. If the enquiry is correctly carried out, and the results appreciated, this can dramatically improve the efficiency and economy of the company involved. It will be carried out to identify

exactly how and why the failure occurred, then to look at the procedures, materials, design, etc., which could ensure that it does not recur.

- **Spin-off benefits**. The results of failure investigation often open up related areas of improvement in design, systems, materials and/or production. No more so than when the investigation includes materials and components which have *not* failed, to identify how they could be applied elsewhere or bring benefits of cost or weight reduction, increasing wear resistance, eliminating stress raisers, etc. It is this philosophy of looking at why components do *not* fail that has placed the aircraft engine industry well ahead of the rest of industry in efficiency.
- **To confuse**. There are sadly examples of investigations where the terms of reference ensure that the resulting report will be technically confusing. This could mean that those initiating it were aware that any readily understood conclusion might point a finger in their direction. While the person or organization instructing and paying must have some say in an investigation, there must be no doubt that the investigator should have an integrity equal to his or her technical ability.

THE STAGES OF INVESTIGATION AND REPORTING

In any enquiry there should be an initial diagnosis. This is when a problem is identified and it is realized that no immediate cure is available. This book discusses the importance of ensuring that the problem is correctly defined before any follow-up action is taken, which might otherwise destroy valuable evidence, or produce a 'shoal of red herrings'.

I give information on the techniques of 'problem solving', discussing how the doctor, detective, accountant, engineer, chemist and metallurgist all use the same faculties of hearing, sight, touch and smell in helping to define the situation, and then consider how the problem can be investigated. The surgeon should not cut, the detective arrest, the accountant bankrupt, the engineer replace, the chemist dissolve, or the metallurgist hacksaw until there is a definition of a failure mechanism to investigate.

The book then discusses the product, that is the Technical Report supplied to the client. It argues that this should have a well defined structure, written with an awareness that the recipient is not a metallurgist, and whilst it should be efficiently presented, it should not be a sales or marketing document. Examples of good and poor reports are supplied.

There is then a section on what is, without doubt, the single most important part of any metallurgical investigation – visual examination. This suggests that visual and macro-examination along with non-destructive testing (NDT) in the vast majority of cases will give the experienced metallurgist a very clear indication of the mechanism of failure. That is, 'How' the failure occurred. This can be quite distinct from 'Why' the failure initiated, which might not be identified until further work and discussion are carried out.

The sections on visual, macro and non-destructive testing describe the techniques applied, their advantages and limitations, then list, describe and discusses the six Mechanisms of Failure. These are:

- Tensile (fatigue, bend)
- Compression
- Shear (torque)
- Corrosion/contamination
- Abrasion
- Thermal

I make clear that in the majority of metallurgical investigations the investigator should have either positively identified the failure mechanism, or have some clear ideas from the visual, macro and NDT information – that is the 'How' – before any destructive or non-repeatable work is allowed.

On the occasions when fracture faces have been subjected to severe mechanical or corrosion damage, there may be the requirement to plan further work such as SEM (Scanning Electron Microscope) investigation, or to carry out random micro-examination, mechanical tests or chemical analysis. However, this will be most unusual as it is very seldom that cracks cannot be opened, or fracture faces cleaned to give some positive clues as to whether the mechanism is tensile, compression, shear (torque), corrosion, abrasion or thermal.

Once the failure mechanism has been identified the metallurgist will want to examine the structure. It cannot be over-emphasized that the 'How' part of investigation need not give the reason for failure. Too many reports assume the mechanism of failure is the reason for the failure. A subsequent chapter therefore discusses the methods used to identify the 'Why', that is the true reason for the failure.

The section on micro-examination discusses how the metallurgist chooses the areas for micro-examination, and describes the techniques used to cut and prepare specimens. It points out that the metallurgist is looking at minute areas and that what is seen at ×150 magnification is quite different from ×500 and ×1000 magnification. This means that there is considerable skill in choosing the correct location, with random specimens seldom resulting in the critical area being examined.

It could be relevant at this point to define a metallurgist. Whereas a metal can be simply defined as an element which, when heated, will show a reduction in electrical conductivity, or in terms of its thermal, magnetical or physical properties, a metallurgist has to accept they will be confused with metrologists and even meteorologists!

On an even less serious plane is the definition of a metallurgist as the person who can differentiate a common ore from virgin metal, and there's the story that the difference between a metallurgist and a psychiatrist is the letter 'n' – one deals with metal fatigue, the other mental fatigue!

But generally speaking, a metallurgist is someone trained to understand how metals are produced and what happens when they are processed or otherwise worked on. It could also be expected that the metallurgist should recognize metal compositions and understand their significance, also the relevance of mechanical properties and how they are estimated, and while this may not extend to being able to calculate stress or strains, he or she should be able to understand the answers.

There is the oft-repeated statement that when two metallurgists are present, there will be at least three opinions – generally for perfectly understandable reasons, if they are allowed to explain!

Following the section on micro-examination there is a discussion on hardness testing, which points out this can more readily be compared to a wooden rule than an engineer's micrometer, which does not mean that the results are useless.

Different diagnostic techniques are described and evaluated, and it is suggested that in conjunction with visual, macro and NDT information the investigator can often form a firm conclusion early on regarding the reason for failure – the 'Why'.

I then highlight the importance of the metallurgist in subsequent failure investigation. There must be the ability firstly to see the mechanism of failure; then to correctly choose the locations for micro-examination; to supervise the correct preparation; to identify the metallurgical structure or structures which appear; then to have hardness tests correctly carried out.

Mechanical tests are then described and discussed, with the reason for carrying out any test evaluated. It is stressed that no test should be called for which does not contribute positively to the investigation of a failure.

The following section is on chemical analysis. This gives a brief note on the techniques available, then lists the 38 elements (out of the known total of 103) which the author considers to be involved in industrial materials. These are then discussed in alphabetical order. Information is given on when the author considers chemical analysis would be justified.

There then follows a brief description of electron microscopy which includes criticism of the rather common use of this sophisticated equipment in lieu of visual, macro and standard optical microscopes. I argue that there is a place for this technique in failure investigation but this will be to enlarge on the findings of conventional tools, rather than as an alternative.

REFERENCE INFORMATION AND CASE HISTORIES

Next is information on material properties and specifications. The technical investigator must be aware of the reason a material is chosen. This section takes each of the metals and non-metals used in industry and gives information on their properties, history and common names, with relevant specifications.

There is then a discussion on how they might fail, what the investigator might look for, and the technique or techniques used to find why failure occurred. There must of necessity be some duplication within this section of information already supplied, but it is considered that this is preferable to constantly referring to other sections.

The metallurgist is very much aware that it is the heat treatment of alloys and steel which allows modern engineering components to carry out operations which would otherwise be impossible. Many failures are the result of incorrect heat treatment, or welding of heat treatable metals. It was therefore felt it would be useful to describe heat treatment in a non-technical manner and how the processes can go wrong, resulting in failure.

The remainder of the book consists of case histories and a cross-index. The case histories are divided into seven categories by failure mechanism, that is: Tensile, Compression, Shear (torque), Corrosion (contamination), Abrasion, Thermal and general. These are brief notes on specific failures, the majority investigated by the author, but including some dramatic national disasters. Each lists the mechanism, any interesting feature, how the reason for the failure was identified, and in some cases illustrates the failure and its consequences. An attempt has been made to include a sample covering each material used in industry.

The initial diagnosis

2

When a problem arises it will be identified by poor performance, noise, a funny smell, appearance of smoke, unscheduled stopping of moving parts, fracture, or some other obvious effect, by which time it is necessary to consider the possible reason or reasons. The first, and without doubt most important, aspect of problem solving is to define the problem! There are volumes of evidence to show that considerable energy and financial resources are wasted in expensive time-consuming investigations which go down the wrong track.

A dramatic example of a wrong failure definition is the two-engine plane which developed a fault in one engine, but the second engine was closed down, resulting in a fatal accident. In this incident the pilots were supplied with less than perfect information from their instruments, and had only minutes to identify the problem. The pilots did correctly evaluate the situation, but were unable to re-start the good engine in time to prevent the crash. In reality, the time scale involved could probably be measured in seconds between failure and success in this instance.

As another example, the tanker oil leakage in Alaska took place over a much longer time scale. Here navigational errors were compounded by not defining the problem over a time scale of hours, resulting in a major disaster, with no human loss of life but very considerable financial loss with serious environmental implications.

TIME TO LOOK, LISTEN AND LEARN

In many faults or failures there is time to consider, examine the situation, look at the parts, and discuss with experts before any dramatic occurrence will arise. This is where experienced investigators are worth their weight in gold as they will question, listen, look, smell, touch, and consider the evidence, with no rapid decision resulting in dramatic action which could destroy evidence or cause further damage through wrong conclusions.

The AA, RAC and similar organizations are past masters in the art of listening to verbal evidence – very often of doubtful authenticity – then looking, listening, touching and smelling before taking any action on a faulty vehicle. The action taken will be based on evidence available, and result in either a solution, or towing to a garage. In such circumstances, it is very rare that a car is towed away, only to be started immediately by the garage mechanic.

There are several quite different techniques used in problem solving, but all invariably commence by ensuring that the 'problem' is correctly defined. No action should be taken until all involved are quite certain that a correct definition has been agreed, or that some work must be carried out. This will be visual examination, discussion with operators and managers, examination of specifications and drawings, and/or evaluation of the operating conditions.

Once the problem has been defined and agreed, then the options which could be used to solve, or investigate, are listed. No matter how simple or complicated a failure, the competent investigator will list a minimum of six ideas for further consideration.

This portion of the exercise involves 'brainstorming'. It is of vital importance that all the possible reasons for failure are listed for evaluation. There should be no rejection of a suggestion until the end of the brainstorming session.

For example, if our car is in a ditch, the problem would be defined as 'How to return the car to the road'. It would not be so much to do with the steering or brakes having failed, that some idiot had not dipped his or her headlights, that it was dark, raining or icy at the time, or that the girlfriend/wife/children are cold and wet, etc. The problem of the moment is how to shift the car from the ditch.

The options could be:

- Abandon in ditch and go home
- Send for a helicopter
- Try to drive it out
- Jack up front
- Jack up rear
- Send for expanding bags
- Obtain planks
- Send for AA, RAC, etc.
- Find six willing, fit, heavyweights to lift
- Find JCB to dig it out
- Pray for help
- Find stones to place under drive wheels
- Stop passing car/truck and ask for a tow
- Use the adjacent tree with a pulley to lift car

It is now accepted that to be a competent problem solver – that is capable of diagnosing and then managing the solution – it is necessary to be able to list, mentally or in writing, at least six possible actions.

It could be the unlikely idea of using a helicopter that rationally leads to the suggestion of using the readily adjacent tree as the fulcrum to hoist the car out of the ditch. The individual who has not the imagination to consider the helicopter as a lifting device will not be a good problem solver.

So, too, with industrial failures and problems, once the ideas or options have been listed they must be evaluated, and only then should work be instructed.

The diagnosis of any failure will make use of certain mental and visual characteristics, and the investigator must have specific technical ability of a high order, with an eye for seeing details, and then the critical facility to evaulate what has been seen.

The first faculty used, however, will generally be hearing. The doctor listens with his stethoscope, the driver of the car hears grinding noises or squeals, the engineer hears sounds of seizure or knocking of the engine. And all disciplines listen to 'explanations' from the client or patient.

The second, and without doubt most important, faculty will be sight. The doctor looks at a rash, the driver looks at his petrol or oil gauge, dipstick, etc., the garage man takes the cylinder head off and looks for problems inside the engine. The detective looks for clues before touching and generally before interviewing any suspect. It is most commonly sight which will find the broken part requiring further investigation. Visual examination is the subject of a separate section in this book.

Another important faculty is touch. The doctor will touch various parts of the body feeling for lumps, hot spots, etc., the engineer will assess the temperature of the engine by touching components which will give some indication of whether or not the oil is low, bearings hot, and any vibration felt will tell its own story. Surface finish can be estimated by the 'thumbnail test, that is sliding the perpendicular thumb across the surface, with the nail in contact with the surface. (This can give quite an accurate assessment of different surface finishes.)

Vibration can be felt with fingers in contact with a surface. A more sensitive assessment can be obtained using a thin metal bar with one end in contact with the unit and the other end pressed against an ear, the forehead or even teeth. This will pick up low intensity vibration.

Smell can be important, particularly at the early stages of mechanical failure. The odour of over-heated oil, burning rubber, sulphide or other chemicals can supply valuable clues to the investigator with the 'educated nose' – most significantly when investigating corrosion problems. Site investigation where hydrogen sulphide, sulphur dioxide or other odours are evident, would also be significant.

Corrosion products should be sniffed carefully in the dry state, and a small sample moistened with water and sniffed again. Taste is seldom used in mechanical problem solving.

Very few organizations have or require the facilities for immediate access to specialist project engineers, metallurgists and managers when problems arise. Exceptions here are the nuclear, aircraft and major petrochemical companies where almost invariably there will be a team on stand-by for immediate reaction when a failure occurs.

Problems generally arise therefore when only semi-technical or non-technical persons are present. We are all familiar with the poor soul standing beside a car, hoping that the problem will go away or be solved by some 'knight' stopping who is capable of diagnosing and curing the problem.

When either the roadside expert turns up, or the car is at the garage, then the first thing will be that the driver is quizzed regarding what happened before the car stopped. This is a verbal discussion regarding noises heard, reactions of the car, etc. The mechanic or engineer will then examine the car in more detail and using experience, decide on the next stage.

There is on record the investigator who diagnosed a car noise as a gear box problem, and stripped and examined the gear box, only to find the noise persisted after re-building and then identified a loose roof rack as the source of the noise.

It is, however, a pleasure to watch the efficient investigator examining a problem which does not have an immediate answer, to see the logical steps taken, each step either identifying the problem or supplying positive information for the next action. This is where the problem solving technique briefly described above is seen to be working efficiently. No action is taken until the problem has been defined.

Having obtained a definition a mental or written list is prepared of the possible techniques which could be used to identify the reason. A written list is recommended which can then be critically evaluated, and the chosen techniques followed. While this will not always cure the problem, it will result in it being tackled in a sensible manner step by step, with each step taking the investigation a stage forward. There will be no redundant work and, therefore, the investigation will be carried out in an efficient and economical manner.

The same procedures should be followed by the operator in a machine shop within maufacturing premises. Whether this is an engineering organization making components, a bottling plant processing milk, beer or lemonade, or a textile factory making stockings, cloth or sewing suits, there will be failure with equipment and there should be an engineer either on site or on call who will first of all talk to the operator, and then make his or her diagnosis, using the techniques described above.

With modern motor cars, machine tools and equipment used in industry, where moving parts are involved, these are now extremely sophisticated and much more reliable with longer life between problems and a higher output. Much of this increase in productivity and performance can be credited to planned maintenance. Both planned maintenance and increased reliability are based on intelligent problem solving. That is, looking at failures which have occurred, examining these in sufficient detail to understand the reason for the failure, applying the correct cure and, where necessary, applying monitoring and servicing procedures.

Failure investigation is a relatively modern science and is still not fully understood and applied. Too often failures are examined without the necessary competence and the replacement or the 'cure' is not the result of any intelligent thinking but merely the first thing that comes to mind or is available to try. Broken items are frequently discarded and replaced without investigation or comment.

Where the component has been under-designed, botched attempts to cure the problem will appear to be successful but will seldom be effective.

There are numerous examples, some given as case histories, where there is a history of problems or failures which are not 'solved', or the conclusion of the failure report is not appreciated or accepted. This can result in considerable financial cost directly as the result of the problem, but quite often because the 'cure' does not attack the problem, or the conclusion is ignored.

One example of this is an offshore oil rig anchor chain where failures were not fully investigated, the recommended 'cure' being to make 'stronger chain'. This stronger chain was manufactured with the same techniques as the original chain, and failures continued until correct failure investigation identified that manufacturing defects were the prime cause of failure, not inadequate strength.

Secondly there is the example of stress-corrosion cracking, now arguably the most common mechanism/reason for failure. This is the combination of stress and corrosion where neither by itself would cause a problem. Copper from copper-loaded grease, or some other source in contact with moist steel can initiate stress-corrosion cracking. Despite this, copper-based grease is still commonly recommended to prevent galling, or reduce friction. The same low friction result can be achieved using zinc, lead-based grease, or molybdenum disulphide without any corrosion problem.

Thus far there has been no mention of a metallurgist or metallurgical investigation. Often the investigator could be a competent engineer or someone who has managed to identify the problem either correctly or with sufficient acumen in order to start the machinery again and satisfy those involved. It is when a process will not start and on further examination something is discovered which is cracked, fractured, bent, worn, or damaged with no obvious cause that the metallurgist is sent for. The role of the metallurgist will be discussed in the remainder of this book.

In the case of the nuclear, aircraft and petrochemical industries, the metallurgist is often involved at the initial stage. There will be a stand-by team in the aircraft and nuclear industries, and commonly in major petrochemical companies, where when a problem of any seriousness is found there will be an on-site metallurgist and an engineer.

The techniques described above will be used, involving sight, sound, smell, touch, and sometimes taste, and these will tell the engineer one story, and may tell the metallurgist another story. There will be discussions and in the well organized team a problem-solving exercise will be carried out, with a manager who may be non-technical.

This team will visit the site, making a visual examination without disturbing the evidence, and discuss the situation with those immediately involved and will then retire to a conference room where there will be a discussion with a team leader, and the various tasks will be allocated. These will be a detailed visual examination, discussion with those involved, and recovery of the relevant pieces for further examination.

In the case of an aircraft accident or in petrochemicals where some form of explosion is involved there is often considerable damage. There are now sophisticated skills available which will be used by the investigator along with experts in metallurgy, chemistry, etc. But whatever the level or depth of the failure investigation it can be carried out at minimum cost if good communication is established between those handling the equipment and the investigator.

REPORTING AND DISCUSSION OF INITIAL FINDINGS

Discussion round the table will on occasion obviate the need for further investigation as the reason for the problem will be identified.

Reporting should be encouraged and where reports are produced, these must be concise and readable. There is a chapter in this book devoted to report writing as it would seem that an investigation is pointless if the outcome is not understood and documented.

The investigator should take notes during the preliminary discussion and use these notes to formulate a plan of campaign. At the end of this initial meeting, when a relatively cursory visual exmaination of the failure has been carried out, there should be a definition and a plan of campaign produced. This could be to partially or fully strip the components, the engine or machine, to identify which areas have to be accurately inspected or measured. Steps would be taken to obtain documentation regarding the design standard of the components involved.

The investigator and client should be aware of the costs which are involved in failure investigation. There seems little point in carrying out expensive operations such as high quality surface finish investigations and accurate dimensional measurements if either the information for comparative purposes is not available or the information itself is irrelevant.

Careful details of the visual appearance of the equipment in the critical areas should be listed and recorded. The actions taken and the reasons for them should be recorded.

Photography can be useful, but care is necessary to identify the photograph, and location details. These are useful for confirming details after the component has been moved, or altered by cleaning. Photography must not be confused with, or used as, an alternative to visual examination.

It is not possible to produce pro forma information to cover all eventualities but the pre-investigation meeting should have routinely highlighted the areas to be examined and the order in which they require to be examined, and these should be methodically checked and the results recorded. If the reason for the problem has been identified then in the majority of cases the only action required will be remedial.

It may, however, be that during the investigation there has been some evidence uncovered that the problem might exist not only with the unit being examined, but with other similar units. This must be recorded and communicated as rapidly as possible in order that a plan can be prepared for further investigation of those units, which need not wait until the full investigation has been completed.

INVOLVING THE SPECIALIST

As the investigation proceeds it will be commonly found that evidence is produced which the engineer or investigator is not competent to understand. At this stage great care must be taken that the evidence for the specialist is not distorted or destroyed. The specialist areas would include metallurgical, electrical, chemical, or other specialities depending on the information which has been uncovered.

It is essential that the evidence is supplied to the specialist without having been tampered with. Very often important information is contained in the debris associated with a failure and if this is removed by cleaning up, then the investigation can be seriously jeopardized. This is especially so for any metallurgical examination but would apply equally to chemical investigation where cross-contamination from handling can lead to false interpretation during chemical analysis. Similar considerations apply to electrical problems.

It is desirable that the specialist is contacted as soon as possible and brought up to date with the investigation already carried out. Strictly speaking another problem-solving session should be held and the evidence produced examined by all concerned. If the original problem-solving exercise has been correctly carried out, there will be no false route and no further evidence to produce other than confirmatory evidence.

The specialist, in conjunction with the engineer, could however decide that the definition of the original problem was not correct and in this case the problem-solving exercise would be repeated.

It is very important to increase the efficiency of industry firstly by ensuring that the reason for a failure is properly identified, and then that the techniques required to prevent failure are clearly stated and followed up. It is equally important that products should be constantly monitored in order that improvements can be made on an ongoing basis. The aircraft industry, and in particular the aircraft engine industry, has pioneered the techniques of failure investigation and improvement techniques and has clearly shown that this is an economically viable science. If the same attitude could be adopted in general engineering then the life of moving components would be lengthened by a marked degree. There could also be considerable savings in weight and cost efficiency as well as benefits from using alternative materials.

Report writing

3

Failure investigation can be expensive and unless it has an end result this expense will be wasted. The technical report is an end product of the failure investigation and should be written in clear English, in an unambiguous style which can be understood by the recipients.

The investigator (and report writer if different) must be aware that once published the report becomes the property of the client but with the copyright remaining with the writer. The report can be used for legal purposes, with a considerable time lapse between producing the report, and it being used as evidence. Once the report has been published (and paid for) the client can use it (without alteration) for his own purposes.

FORMAT

Reports should have a format. This applies in particular to metallurgical reports the structure of which should be capable of being interpreted with a system of easy reference. Reports should contain:

- Introduction and background
- Conclusion
- Recommendations
- Visual examination – description of the circumstances of the incident, identification markings, dimensions, etc.
- Technical investigation – non-destructive tests, macro/micro examination, hardness tests, mechanical tests, chemical analysis, etc.
- Discussion

The introduction, conclusion and recommendations should be the initial part of the report, and must be sufficiently complete as to be able to be separated from the remainder of the report. Where it is felt that any one, or all of the three parts, require further qualifications, this can be achieved under 'Discussion' at the end of the report following the technical details.

The recipient of the report should be given a clear picture of the complete situation after reading the first three sections. Either the conclusion or the recommendations may refer the reader to the discussion, but the sections themselves should be kept concise with no ambiguity. Thus a typical metallurgical failure report would proceed as follows.

Introduction and background

This would state the terms of reference between client and investigator which could be one sentence and rarely more than a paragraph defining the client's requirements as understood by the investigator, stating the type of component, and general but specific information on the nature of the problem. It should be short and not involve anything except the information as originally supplied, with relevant background circumstances.

Conclusion

Again, it is essential that this is concise and clear. The conclusion must always be based on the findings in the body of the report and should be positive, not vague, and follow the Introduction.

Where necessary, reference can be made to the 'Discussion' for expansion of the conclusion. A negative, that is 'don't know' conclusion is preferable to a long waffling dissertation which leaves the client to form their own conclusion.

Recommendations

There will be occasions when recommendations would be pointless or for many reasons could not be supplied by the investigator. There may even be an argument for stating at the end of the conclusion that no recommendations are possible, giving the reasons.

However, the purpose of an investigation is to isolate the cause of a problem and the competent investigator should aim to give some form of recommendation, and should certainly be capable of helping prevent further incidents.

The recommendations need not necessarily produce detailed specifications but give some indication of the areas where either further work would be carried out or information on how to proceed to improve the situation. If necessary an appendix can be added giving a detailed specification if within the investigator's remit and experience.

It may be necessary in the process of giving a recommendation to be aware of the commercial and legal consequences and to make certain that the client is aware of any possible implications.

Any recommendation should be short and concise, with reference made to the Discussion when expansion is necessary.

Visual examination

Without doubt this is the most important part of a failure investigation. It should commence by listing markings, then supply dimensions and a concise description of the parts involved and the site details, and followed by the findings of a macro-investigation.

The initial visual examination and identification of the failure symptoms and mechanisms are often the aspects most relevant when legal action is involved. It is therefore vital that the report when read after one or two years by both the writer and the recipient will allow them to accurately recall and trust all the relevant features of the initial examination.

Photography can be helpful in aiding a visual description of the site of failure. It is essential, however, that the written description is fully comprehensive on its own and can be understood. The recipient should not be expected to use a photograph to carry out their own visual examination and corrective action nor be misled by what is highlighted or omitted by the camera.

Technical investigation

In the majority of failure investigations micro-examination is the most important section after visual examination and identification of the failure mechanism. This will identify metallurgical structures which have specific names meaningful to metallurgists. The report must make certain that when these terms are used that their relationship to the reason for failure is clearly identified and explained. The report writer must be careful to refer to the positive metallurgical aspects identified, and not to confuse the recipient with a lecture on metallurgical structures which are not relevant.

Hardness testing will commonly be carried out and reported. The report must state the area where the testing was carried out and the type of testing used and then interpret the results. All specification details of the component under test must be included and the relevance of any difference in hardness from the specification explained.

Mechanical testing may be called for and the reason for having a test carried out must be explained and the results interpreted. Specification information will again be reported and any deviation from this clearly explained. It is the responsibility of the report writer to explain in simple language the relevance of the test, and whether or not any difference to the specification is significant.

A good technical report will not include figures or tables without explanation.

Chemical analysis is often carried out as a routine part of investigation. If a correct problem-solving exercise has been carried out the need for chemical analysis will have been identified and will then be part of the report. The analytical result will be tabulated together with specification information, and these will then be interpreted. Investigators must be aware that reports will be read by non technical personnel who could become confused by information which has no relationship to the failure. The relevance of the results obtained to the failure must be fully explained.

A good report will only have chemical analysis results reported which supply positive information.

Discussion

While not essential, it is often useful that a report includes a section which enlarges on the investigation. It is important that the conclusion is made from facts positively identified during the investigation. No conclusion should be made because of an investigator's opinion alone, based on facts found elsewhere which might exist here but were not actually found at the investigation. The discussion can bring out opinions or ideas while enlarging on the factual aspects of the report such as micro-examination and hardness testing, with care being taken that the discussion does not support the conclusion in a manner which could confuse or leave a gap in the explanation. It may suggest that further work be carried out, and this could lead to a meeting with the client to discuss the report and how additional investigation would be useful, in which case there should be some mention of this suggestion in the recommendations.

Technical jargon

The majority of reports will include technical names, to describe metallurgical structures, types of fracture and/or the units used to measure dimensions or strength. It must not be assumed that readers will know the meaning of a word or even a chemical symbol. Acronyms and abbreviations can cause confusion and should never be used without explanation.

Reports should therefore not in general use abbreviations and should always make certain that there is a definition of technical terms in simple English.

It would appear sensible that all UK and European reports make use of SI (Système International) units, but make certain that units familiar to the client are also included.

Thus, for example, a report for an American-based client with mechanical test results would show SI and Imperial units:

Sample	*Tensile strength*		*Yield strength*		*Impact*	
	N mm^{-2}	lb/in^2	N mm^{-2}	lb/in^2	J (0°C)	ft lb (32°F)
A	540	78 340	480	69 640	40	30

For a UK client the following is suggested:

Sample	*Tensile strength*		*Yield strength*		*Impact*
	N mm^{-2}	tons/in^2	N mm^{-2}	tons/in^2	J (0°C)
B	900	58	750	48.5	25

For some clients it may be necessary to show that N/mm^2 is newtons per square millimetre, lb/in^2 is pounds force per square inch, etc. The units in which the machine is calibrated would normally be quoted first. This is discussed in more detail in Chapter 10 on Mechanical testing.

The report writer should take into account the technical knowledge of the likely recipients, and if any doubt exists no jargon or abbreviations should be used without a full translation being given. It can be extremely frustrating for the client to be presented with figures and text which are quite incomprehensible because of the poor use of abbreviations or terminology. It can be very difficult for a non-technical person to find a practical translation.

Layout and illustration

A report should look attractive. A single page report which is a mass of close print can be offputting whilst a well spaced two-page document will be more easily read.

Headings when used should be distinctive. There should be ample space between paragraphs and adequate margins, these being useful for note making during reading.

Paragraphs may be numbered for cross-reference, but care is required that these remain accurate after editing, and do not detract from the visual appearance. Many report writers do not use a numbering system.

Care is required in the use of illustrations. The reasons for this are outlined in Chapter 4 on Visual examination. There is no doubt that they can enhance the appearance of a report, but this is a technical document supplying information, and need not be artistic. While there is no harm in using a quality document to enhance the reputation of the report writer, it is important that the visual appearance does not become more important than the technical content.

Illustrations of all types should be used only to include necessary information, and wherever possible be associated with the relevant text.

Where tables and charts are included they must be clearly laid out, with all abbreviations defined, and translations of terms readily available.

The charts should be part of the relevant section of a report, where the information can be quickly referred to. Most readers object to having to refer to appendices for information which should be in the report, unless of indirect interest or extensive in detail.

Editing

The best reports need time to mature! Reports may have sections written by different people at different times. These should be read and edited by one person with edited drafts either left for a period of time then re-edited or given to an independent editor/proof-reader. Any word or phrase which can be removed or simplified without altering the technical sense should be deleted or changed.

Most recipients of reports are busy people who will want to skim read first, then return to relevant sections, perhaps to clarify, query, or highlight some point. Commonly comments will be made either to critcise the report, or to request further information; hence the need for good layout with white space and margins.

SAMPLE REPORTS

Following are examples of reports. They are two reports on the same problem – one along the lines suggested in this section, the second a more academic report which identifies the problem, but presents the information in an elaborate and at times misleading form.

There is also the case of a massive report issued following an expensive failure. This had a large circulation, no-one however could understand the findings. The author of this book was contracted to interpret the findings and produce a report in line with the ideas listed above.

The first report is 425 pages long without conclusion, and was probably designed to impress or confuse. The author's report with introduction, conclusion, etc. was 4 pages in total! See Figure 3.1.

Being involved in helping clients to interpret such reports was one of the reasons for writing this book!

Figure 3.1 A 425-page failure investigation report reduced to four pages by the author.

Report No. XYZ
Investigation of
Leaking Heat Exchanger

Dear Mr Bloggs

You have heat exchanger units with cast aluminium coils. These are installed in the hospital boiler house.

Some units were found to be leaking, and one such unit was submitted for examination as corrosion was considered to be a possible cause.

This was identified as Model X, Serial No. 12345.

Investigation showed that certain types of heat exchangers which had aluminium cast coils were leaking apparently because of corrosion. One unit was submitted for investigation and report, where leakage was found at one end.

The investigation showed that no corrosion existed on this particular unit but that the gasket at one end had failed. Failure would appear to be because of faulty fitting of the gasket member.

It was suggested that units of this type should have this area examined if the problem is considered to be serious. Visual examination showed that the unit was cylinder shaped with a height/length of approximately 800 mm and a diameter of 400 mm. The unit had steel top and bottom plates and approximately 40 aluminium coils present running around the other diameter. Leakage was identified at one end and the other end was open for inspection. Using a Boroscope the inside of the tubes was examined and while some pitting of a slight extent was identified locally, this was not sufficient to result in leakage.

The unit was therefore placed in a vertical position and the tubes were filled with water. These were left in situ for some time and it was found that a limited number of the tubes showed a reduction in water level.

Examination revealed that the area where these tubes were located on the underside was where the leakage existed. Close examination then showed that leakage was associated with the gasket and when the seal was broken, it could be seen that the gasket was trapped and that this was the source of the leakage.

At this stage you were contacted and it was agreed that no further work was required.

Yours faithfully
for Mistral Technology

John Smith

Report No. XYZ
Investigation into Leakage Problems
on Aluminium finned tubed Heat Exchangers
fitted to a Boiler – Model A/B/C
Installed in a Hospital

INTRODUCTION

We were requested by you to establish the cause of the leakage problems currently being experienced in the heat exchangers installed in the boiler house of a hospital.

Additionally, the client sought information that corrosion of the aluminium finned tubes, if present, was not severe enough to contribute in any way to the leakage problem either directly or indirectly.

BACKGROUND

(1) Since the time of the boiler house in the hospital coming 'on-stream' in 1988, the Plant Maintenance Dept has been experiencing recurring water leakage problems in the heat exchangers at the top cover gasket area.

(2) As the heat exchanger is constructed in aluminium, chemical treatment of the water flowing through the finned tubes is necessary to avoid

corrosive attack on the material. To be effective, such treatment must of necessity be ongoing throughout the design life of the component.
(3) Information obtained from plant personnel however, indicates that until 1990 such treatment was inconsistent in its application, and it was not until the latter half of that year that steps were taken to regulate the chemical treatment of the water through systematic analysis at agreed intervals. This task was delegated to the local representative of XYZ on a contractual basis, and is currently in force.
(4) On several of the occasions when leaks have occurred in the heat exchangers, the manufacturer's service engineer has substituted different gasket materials in an effort to obtain a permanent fix to the leakage problem. It is relevant at this point to record that specified torque values for the top cover panel securement bolts are laid down in the 'Users Instructions' document supplied by the boiler manufacturer.

INVESTIGATION

(5) Subsequent to visiting the boiler house to examine at first-hand the installation, and observe evidence of water staining present on the front cover panel at the heat exchanger positions, a complete unit was removed to the laboratory at Sirocco Materials for examination. This consisted of Items 11 (plus 20), and 12 through to 18 as illustrated in Fig 1.
(6) In essence the investigation was conducted in two parts. One concentrated on establishing whether or not localized distortion of the gasket face of the aluminium top tube plate (Item 13 in Fig. 1) was a contributory factor in causing leakage always to be evident between the 5 o'clock and 6 o'clock positions when viewed from the front of the boiler module.

The other part of the investigations involved visual and metallurgical examination of the aluminium finned tubes to identify whether or not corrosion was present, and if so, to what extent.
(7) Dimensional checks: The residue of the gasket material adhering to the adjoining faces at the cast iron top cover and the aluminium top tube plate flange area was removed for flatness checks on both items.
(8) With respect to the cast iron top cover, this face was found to be substantially flat. However, in the case of the aluminium top tube plate, a straight edge positioned across the outer face diameter revealed that the central joint face area was reading 'low' to an extent of 0.13 mm/0.38 mm (0.005″/0.015″).
(9) Additionally, it was noted that the central nine (9) M8 studs were held in position on the central area of the aluminium top tube plate by helical inserts. Contrary to accepted engineering practice however, some of these inserts were standing proud of the central joint area; it is normal practice to drive the tang-ends of the inserts below the level of a joint face.

Examination for Corrosion

(10) It was apparent from visual examinaton that in the case of the unit in the possession of the laboratory, extensive leakage had occurred past the central area of the gasket. The water had traced a path down the inside face of the top tube plate, flowed through the interstices of the finned tube stack, and dropped on to the gas distribution screen (Item 24 in Fig. 1). It evaporated in these hot zones and left a residue of solids on the inside face of the gas distribution screen.
(11) As indicated previously, water staining was also present in the 5 o'clock/6 o'clock positions on the outer joint faces in both the cast iron top cover plate and the aluminium top tube plate.
(12) With no evidence of corrosive attack at the top end of the heat exchanger, the bottom cover plate of the heat exchanger (Item 18 in Fig. 1) was removed. No indication of leakage at this joint was found, nor was any trace of corrosion present on the aluminium bottom tube plate.
(13) By use of an endoscope instrument, the bore of every finned tube in the stack was examined for the presence of corrosive attack. No significant corrosion was found in any of the tubes, nor were any leakage paths found in those regions of the heat exchanger.

Some of the tubes revealed superficial/light corrosion products which were capable of being displaced by fibre brushing the bores of the tubes. No evidence of corrosive pitting was found in the tube bores.

CONCLUSION

(14) From the evidence obtained from examination of an aluminium type heat exchanger removed from a modular gas fired boiler, it is considered that leakage problems being currently experienced have their origins in the mechanical assembly of these units.
(15) No evidence of primary corrosion was found to be present in the heat exchanger. Thus corrosion played no contributory role in the leakage problem.
(16) The lack of flatness across the entire face of the gasket area found in the aluminium top tube plate, combined with the tang-ends of helical inserts standing proud of the gasket face area, contributed to pronounced water leakage in the central gasket area. The end result was ingress of water into the hot zone region of the heat exchanger.
(17) The visual evidence of leakage at the outer perimeter of the cast iron top cover/aluminium top tube plate gasket joint is considered to arise from localized gapping of the joint.

RECOMMENDATIONS

(18) The possibility of reducing the overall central gasket area, combined with restoring flatness in the aluminium top tube plate gasket face, in order to increase the clamping force in that zone and hence overcome the leakage problem present in that region to be addressed to the boiler manufacturer.
(19) The boiler manufacturer also to address the leakage problem at the outer perimeter of the gasket joint and determine whether or not movement of the water return elbow component is responsible for localized gapping of the gasket joint.
(20) The frequency of chemical analysis and allied water treatment for the heat exchangers be increased from the current quarterly intervals to once a month. This is to obviate against possible excursions of the closed water system into high alkalinity levels going undetected for long periods.
(21) Based on the evidence found after examination of the heat exchanger supplied to the laboratory, the cause of the failure, through water leakage through the gasket area, points to it being mechanical in origin. The absence of any pronounced attack on the aluminium material rules out corrosion as a contributory factor to the leakage problem.
(22) With respect to the central gasket area on the top tube plate, it is considered that it displays a 'weak area' in design in that it contains a relatively thin area of gasket contact adjacent to large areas of gasket contact.
(23) In our opinion, it would probably be advantageous to locally scallop away some of the 'large' contact areas, thereby creating a reduction in the overall contact area of that region of the gasket. In effect, we would be increasing the 'clamping force' in that zone since the M8 screws would be torque tightened over a smaller surface area.
(24) If the lack of flatness over the entire gasket contact area of the aluminium top cover plate (see Para. 10) were corrected by machining, then this feature when combined with the suggestion mooted in Para. 11 might eliminate the leakage situation across the central gasket area of the aluminium top tube plate. This applies to the heat exchanger submitted.
(25) It may well be that the 'lack of flatness' feature is present in other heat exchanger units in the boiler house installation. In view of recurring leakage problems, experienced by the plant personnel at the hospital, this is a reasonable assumption.
(26) With respect to the leakage which manifests itself visually (locally) at the outer gasket contact area of the cast iron top cover/aluminium top tube plate joint, it is considered that the position occupied by the water return elbow (Item 22 in Fig. 1) is relevant to explain the localized leakage at the 5 o'clock/6 o'clock positions.

(27) If vibrating forces are being transmitted via the return water manifold to this elbow component, then locally the M10 bolts are being exposed to additional tensile forces on top of the applied tension due to torquing in the bolts, and thus localized gapping at the joint, hence leakage would result in this area.

Yours faithfully
SIROCCO Technology Ltd
John Brown

PART TWO

Failure Detection

Visual examination

4

Without doubt visual examination carried out by a skilled, competent investigator is by far the most important aspect of mechanical failure investigation. It is visual examination that in the main enables the investigator to identify the mechanism of failure.

Visual examination makes use of either a hand-held magnifying glass or binocular magnification in the laboratory in addition to unaided eyesight.

ON-SITE INFORMATION

Wherever possible the initial examination should be carried out on site and allow no preconceived ideas. The problem should be approached with an open mind, resisting all voluntary assistance or information. For example, the author was once witness to an accident where the driver of the bus involved was unaware of the incident and drove on, leaving an injured man. One witness 'saw' a red bus, the other a green and yellow bus. Neither witness had any axe to grind, and buses of both colours used that route!

The visual examination should therefore be carried out before any detailed discussion, particularly discussion with others involved. The investigator must first obtain a general idea and impression of the equipment and make notes of the site, with general dimensions and brief description of the items involved. These need not be in detail but should be sufficient to remind the investigator of the salient points when further investigation is being carried out.

A magnet, knife, torch and a pin can be useful at the site of failure. This examination will look carefully at any fracture or damage and will note the relationship between this and associated equipment.

Depending on the initial findings the investigator might need to examine the whole area and take samples. These might be of liquids or other materials in contact with the problem.

There could be occasions where the atmosphere required examination, and samples taken.

The wise investigator will insist on visiting the site to collect the problem pieces, and after the initial examination to discuss with operators in addition to management what has been seen. It is not unusual for a chance remark to supply vital information, which either supplies the answer, or allows the investigator to home-in on a specific area, thus simplifying the investigation.

A hand-held tape recorder can be usefully employed for these notes. Approximate dimensions of the equipment should be recorded with detailed dimensions of any pieces which seem to be relevant, such as the exact size of a bearing. The investigator must either take these dimensions, or ensure that they are obtained from an independent responsible person.

The investigator must always be aware at this and all subsequent stages that the facts recorded could be used in court several years later. The general state of the equipment must be noted and recorded. Very often evidence is available in the debris, which where necessary should be collected and taken to the laboratory with other components. It can be embarrassing to find that critical evidence had been disposed of before examination had been carried out.

FACTS AND OPINIONS

Once the initial inspection has been completed, which should take a very short period, minutes rather than hours, and the initial indications have been noted, then discussion can take place with others involved. It is important, however, that the initial investigation should be approached without any prior ideas or suggestions.

At this early stage the investigator should form an opinion regarding the mechanism of failure. This might not be firm because of problems with contamination or bad light, etc., but there could be indications, for example, that fatigue is present. This initial opinion might lead to a request for further test pieces to be produced and worked under similar conditions, or for further information on specifications or drawings to be made available.

If components are submitted for an investigation, it is desirable that the components are examined without bias before the written comments are read.

Once the initial examination has been completed, and the investigator has formed an independent initial opinion, only then should the situation be discussed with as many involved as possible, and any written submissions examined.

It is important that the investigator listens very carefully to all the 'facts' which are produced, and not make any judgement. It is not unknown for what is at first sight irrelevant or even impossible information to contain details relevant to the failure. It is quite common for vital information to be given in general conversation with operators or others who are quite unaware of the relevance of the information being supplied.

The investigator must be a good listener and be able to put those involved at ease. During conversation an explanation of what has already been discovered should be given in simple non-technical language. It is important that this discussion takes place without any indication of superiority, with the investigator describing the effects noted, explaining their significance, and indicating the next stage of investigation.

It must always be realized that the initial opinion could be wrong, and when vital relevant facts are produced a readiness to have another look at the problem and eat humble pie is necessary!

If, however, the visual evidence is clear, and the investigator confident, the client should be contacted as it is more than possible that the reason for failure can be identified with discussion alone.

There was the case of a truck-mounted crane where the pedestal had failed. It could be seen to be a fatigue – a tensile failure. The investigator was informed this was not possible, that the heavier the load lifted, the greater the compression at the pedestal. The investigator had sufficient confidence however to confirm that the failure was in fact through tensile fatigue. Further discussion with the operator found that the crane was used for lifting scrap cars on to the truck. The head of the crane was then used to crush the cars, thus pushing where the design allowed only for lifting or pulling.

OFF-SITE EXAMINATION

In general, once the site examination has been completed and notes made, the pieces should be cleaned. Considerable information will be obtained from the fracture faces, firstly in the 'as received' condition but more so after careful cleaning. If there is deep-seated corrosion on either the whole or a portion of the fracture face this gives an indication of the time involved since fracture occurred, which can be related to the conditions under which it operates. A skilled investigator will have a reasonably accurate impression of the time involved since cracking or fracture took place.

This cleaning must be carefully carried out, firstly with detergents and then if necessary using chemicals, for example phosphoric acid for steel, and caustic for aluminium and other materials. All debris and where necessary the resultant solution and cleaning equipment should be retained.

Where a failure has corrosion evidence it is essential to take samples of the corrosion products for chemical analysis. Samples must be placed in clean containers for further examination as cross-contamination by handling can give false results. Once the fracture faces are clean, they should be re-examined, this time with prior knowledge of the original indications but still with an open mind to look for other evidence and a willingness to alter any opinion based on this new evidence. This examination will normally be in the laboratory using binocular magnifying glasses with good lighting and where necessary handling equipment. Where a hand-held magnifying glass is used, good lighting is necessary.

A description of the correct use of a hand-held magnifying glass is given in Chapter 5 on Macro-examination. Normally a hand-held magnifying glass or loupe is used at magnifications up to ×15 at the most, whereas the laboratory bench binocular magnification can be as high as ×50, or even ×100 magnification.

DETERMINING THE FAILURE MECHANISM

By this time the skilled investigator will in most cases have firm information on the mechanism of failure, if not the cause.

Failure mechanisms come under six headings as follows. Visual and other characteristics of these are given in Chapter 7.

- **Tensile stress**: this will include brittle and ductile tensile stress, fatigue and bend. The component has been subjected to a pull.
- **Compression**: the component has been pushed, and will usually be distorted.
- **Shear**: this is a combination of tensile stress and compression and is seldom found except in overload situations. **Torque**: this is a form of shear and is more common. The component is invariably twisted.
- **Corrosion/contamination**: this takes many forms and can often contribute to other modes of failure such as fatigue. Corrosion is a chemical attack of the surface, whereas contamination is additive to the surface, both producing visual effects.
- **Abrasion erosion**: caused by wear. This is a surface effect, thus visual, and can be distinguished from corrosion.
- **Thermal effects**: overheating or excessive cooling, seldom found by visual examination alone.

No action should be taken on failure investigation until the investigator has identified the mechanism of failure, or at least formed an opinion.

When the problem is one of cracking the full extent should be found, and where necessary photographs taken. Only then should the cracks be opened to observe the mechanism of failure. It may be desirable to have the component crack test inspected before any opening to ensure the full extent has been found.

Great care must be taken before opening the crack that all the surfaces and associated areas are examined for stress raisers, corrosion pitting or any other possible source of crack initiation. It goes without saying that it is much easier to identify these before the crack is opened and to note their location accurately.

Once the mechanism has been identified this can be related to the information obtained on site or during discussion with others involved. It is advisable at this stage for contact to be made with the client as isolation of the mechanism of failure may identify the reason for its cause. There may then be no need for further investigation as the knowledgeable person using or responsible for the equipment will be satisfied with a report.

An example of this is a nose wheel on an aircraft where an insert had broken. Visual examination showed that this had been subjected to side bending in both directions resulting in cracking in stages with final failure by tensile stress. When this information was supplied, the operator could immediately identify with other evidence that a certin action had been taken resulting in excessive side forces being applied. All that was then required was a hardness test to confirm that the material was in the approximate tensile range specified for this component.

In many investigations, however, the mechanism is of interest to the client but does not supply sufficient information for the investigator to be confident that the cause has been identified and can be eliminated.

There is growing evidence that many investigators today do not appreciate the importance of visual examination. Too often instructions are given to cut specimens, carry out mechanical tests, chemical analysis and even scanning electron microscopy without having identified the mechanism of failure. There is also evidence that there are investigators who are not trained to 'see'. If a crack cannot be seen, it cannot be investigated! If a fatigue cannot be recognized, then the mechanism has not been identified and without positive indications of the mechanism of failure, reporting can be meaningless or even confusing.

The information supplied by visual and macro-examination with macro and non-destructive testing will lead to certain actions, thus it seems desirable to determine the various failure mechanisms, illustrate these and indicate the type of information made available to the investigator, and what other steps can be taken or advised.

Chapter 7 on Mechanisms of failure lists and describes, with illustrations, what the investigator sees, and how an opinion is formed regarding the mechanism of failure with possible subsequent actions.

It does not, or should not, require a qualified metallurgist to carry out high quality visual examination, but there is little doubt that it requires training by a competent person.

There is a considerable volume of evidence that at the time of writing many investigations are being carried out without this necessary skilled visual examination. The large (425-page) report recorded at the end of Chapter 3 on Report writing included photographs of a failed chain link, showing a brittle fracture on one of the legs of the link, but also showing stretching, that is ductile failure, on the other leg. This ductile stretch must have occurred before fracture and was therefore the prime cause of failure. The report does not mention this critical fact.

The author has seen many such reports and been involved in numerous discussions where critical evidence, sometimes quite obvious, has not been noted by the investigator.

It is common practice to have macro-examination and non-destructive testing carried out in conjunction with visual examination to confirm the mechanism of failure, but it should be rare indeed for the competent investigator to be uninformed regarding the mechanism of failure at this stage.

It should always be possible for training to be given to identify brittle or ductile tensile stresses, fatigue, shear, torque, cracking, corrosion, abrasion and erosion problems. It must be appreciated however, that there are well-trained people who are incapable of 'seeing' significant details on site.

Macro-examination 5

Macro-examination is by definition part of the visual examination and is not generally shown as a separate section in technical reports. Macro means ‘visible to the naked eye’ – large-scale (as against ‘micro’) – a prefix denoting ‘not minute’. Macro-examination uses ×5 to ×25 magnification with perhaps ×75 where the object can remain in one piece. Good lighting is essential with the correct use of the hand-held magnifying glass or ‘loupe’ – the eye-held glass.

This method is used to confirm areas which will require micro-examination, when the metallurgical structure will be seen, and where there is no visual relationship between the structure and the object.

Visual examination will include the use of hand-held magnifying glasses up to approximately ×10 to ×25 maximum, which is essentially a macro examination. To use a hand-held magnifying glass correctly, the lens is held close to the eye, and the object is brought into focus, either by raising the piece to the eye or lowering the face (and eye) to the area being examined. The focal length at ×10 magnification is short and side illumination is essential. A skilled watchmaker, instrument technician, etc. will be seen to place the ‘loupe’ against the eye and hold it there with the facial muscles so that the lens is almost in contact with the eye.

LABORATORY EXAMINATION

With large dirty and sometimes hot items this technique results in contamination of the investigator’s nose, hands and knees. The experienced and practical metallurgical investigator can often be identified by blistered nose and dirty knees!

Macro-examination carried out in the laboratory will be conducted on a prepared specimen where a cross-section slice is cut and a reasonably good finish produced, in the region of 40–80 micro-inches.

This is a measurement of the number of ridges or peaks. The micro-inch measurement is the one commonly used in the UK and USA. The higher the number, the rougher the finish. Up to 10 would be a honed or polished finish, 10–20 a ground finish, with 20 being a relatively poor ground finish. 15–40 would be a good turned finish, 40–80 would be a reasonably rough finish suitable for initial examination. A micro specimen would be better than 2 micro-inches.

It is a measure generally understood by engineers and metallurgists, and is referred to as the 'thumb-nail test'.

The specimen would be examined prior to etching for any obvious defects such as cracks, or poor radii. The investigator will also see welding effects such as porosity, undercut, that is a notch at the edge of the weld cap, or lack of penetration, that is the root of the weld is not flush or is proud of the base metal.

Etching

The specimens can then be etched using similar chemicals to those described for micro-examination, but at a higher concentration (see pages 114 and 119–20). Where micro-examination uses 1% nitric acid in alcohol, the macro-examination will need 5% nitric acid in water or alcohol.

The etch will not show the metallurgical structure but will highlight accurately welds, and effects, such as case hardening, and forging lines.

Flow lines

Flow lines caused by the concentration altering during the cooling cycle results in 'banding' where the different metal constituents occur at different concentrations. When the casting is forged this banding appears as flow lines, for example in a bar, along its length. If the bar or billet is then used to produce a forging the flow lines will follow the shape of the forging. Where the etch shows well-defined flow lines, this could highlight suspect 'through thickness' strength properties. It is accepted that most wood has properties along its length which are superior to those across the section. Thus wood can be readily split with an axe in one direction, but not at right angles. Flow lines in metal result in a similar situation.

Thus, for example, a component fillet welded to a surface of another part, then loaded in tension, could cause problems similar to glueing a tag to a book cover, then pulling the tag which would lift the cover and open the book. Similarly, a fillet welded to a metal panel with well-defined flow lines could separate the surface layer at low loads.

Re-entrant flow lines, that is where the flow lines go back on themselves, are generally undesirable and can result in stress raisers. Where forgings are incorrectly located during machining, the flow lines can become stress raisers on the finished component.

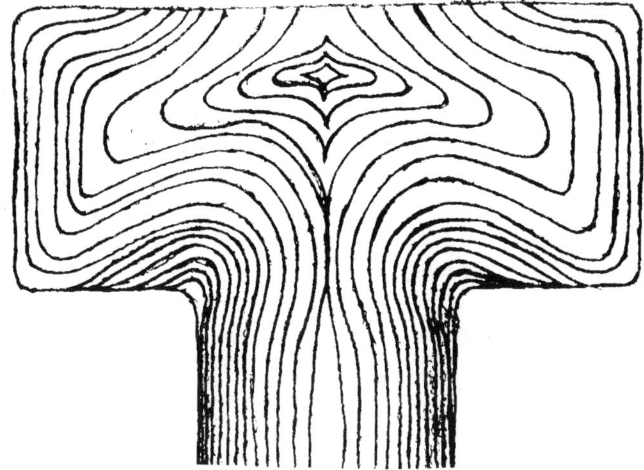

Figure 5.1 Flow line in forging highlighted after sectioning and etching.

Figure 5.1 shows the flow lines when a bar is upset (forged) to produce the head of a bolt or stud. This can initiate fatigue, or in some circumstances result in fibrous fracture.

Micro-examination can supply additional information regarding the different metallurgical structures which can exist with the 'bands', or 'banding' first seen at the macro-examination.

Macro-examination can be divided into three areas of application.

An aid to visual examination

This will use a hand-held magnifying glass up to ×10, and then in the laboratory, binocular magnifying glasses with stereo vision. These can be with fixed magnification, usually starting at ×5 or ×8, going up to ×5 or ×40.

There are now available very efficient zoom binocular magnifying instruments which have continuous magnification from ×5 to ×100. It is doubtful whether magnifications above ×50 are useful for a macro-examination as the detail becomes blurred and the surface finish over-rides the details being examined. With zoom lenses the investigator must always be cautious of what is being viewed and compare this with an item of a known dimension such as a pin.

In the majority of failures, where a fracture is involved, the visual examination will immediately identify brittle, ductile, fatigue, shear or torque modes of failure. The characteristics of these fractures are described in Chapter 7 on Mechanisms of failure.

The investigator will then have to decide whether a macro specimen is required or a micro-specimen/s needs to be prepared.

If the failure mode is clearly identified by the visual, it is unlikely that a macro-specimen would be required. There will, however, be occasions when a macro-sample is necessary to show a cross-section.

With failure examination, details such as fatigue will be visible (as described in Chapter 7) with the starting point identified at magnifications as low as ×10, and seldom need magnifications greater than ×20 or ×40 maximum.

Cracks can normally be identified at ×20 and it is doubtful if cracks of any significance will require magnifications above ×50. Cracks may be present which are not identified at ×50 magnification, but if these contributed to the failure they would be found at the micro-examination.

Indications of stress-corrosion cracking will be found during this examination, showing cracks which bifurcate. These can be very fine cracks, and where galvanic or electrolytic corrosion is the cause, can be very local. They are commonly the stress raiser which initiated fatigue. Although a very significant stress raiser, the extent of the stress-corrosion cracks can be quite small, and readily missed.

There are indications that many more failures and unexplained features are the result of stress-corrosion cracking than is generally appreciated.

Once a section is cut, prepared and etched, then other effects such as confirmation of stress corrosion, welding defects, etc. will be seen, and the investigator will arrange for micro-specimens to be cut, prepared and examined.

Weld investigation

Where there is a weld then details can be seen. Indications of excessive spatter, lack of penetration, undercut or general untidiness can all be readily identified at magnifications of ×10.

It is now common, in fact obligatory, for high quality welds to use procedures which lay down control parameters, and the method of testing. After visual examination the simplest and most economical test is the macro-examination. A cross-section should be obtained through the stop/start region of the weld and through the weld run. This will show fusion, lack of penetration, undercutting, porosity, slag inclusions and/or shrinkage cracks.

Fusion is shown as a mingling of the weld with the base metal. Lack of fusion is seen as a well defined line, or maybe crack.

Undercut and lack of penetration are described in Chapter 10 on Mechanical testing. Undercut appears at the toe of the weld cap as a notch. Lack of penetration is when the root run is not flush.

Slag inclusions are non-metallic particles trapped in the weld, and shrinkage, also referred to as porosity, is holes or cracks.

Standards exist regarding the degree of undercut which will be accepted, the length of each leg of a fillet weld, and the accepted standard for porosity. Cracking of any type identifiable at ×10, ×20 magnification will not normally be acceptable. Further information will be found in Chapter 10 on Mechanical testing.

Where failure investigation has identified any of the above as a reason for the problem, then the report should advise that test pieces should continue to be produced prior to production and if necessary during production.

Raw material acceptance

Where high quality components in the form of bar, casting or forging are specified then macro-examination will be part of the acceptance standard. This will be on prepared specimens at magnifications up to ×20 magnification. The specimens will be examined unetched, and then etched for distribution of flow lines, and evidence of segregation, oxide or slag. Oxide and slag appear as non-metallic particles in the casting or at the surface, while porosity exists as holes in the cast structure. Porosity will generally be more acceptable than oxide or slag, other things being equal.

The specifications may lay down the amount of inclusions in a specific area and also the maximum size, length or width. That is, the design specification for the component could clearly state the maximum size of a non-metallic inclusion and the number of such inclusions which would be acceptable.

With high quality castings a trial casting will be produced under controlled conditions and will be cut and macros prepared in several directions. These will be examined in the etched and unetched condition for the presence of oxide, or slag content, and porosity.

Where macro-examination formed part of the original material specification on a component which has subsequently failed in service, the original findings will be helpful and should be made available to the failure investigator, who will then be able to identify the design requirements relative to the facts which have been identified at the failure examination. The investigator can compare the results from failed specimens with the specification requirements and visual samples.

With non-metallic components the macro-examination very often will be more useful than a micro-examination. With composite materials, the carefully chosen macro-samples will identify the relationship between the different materials involved. This will show the ratio of different constituents, such as glass/carbon/metal fibres to the matrix. It will identify the diameter of the fibres, and whether or not these are homogeneous and within specification. Lack of adhesion, porosity, problems with the lay and make-up will be identified.

Concrete, stone and brick macro-samples taken from various areas can show whether or not the material is homogeneous. The specimens will also show porosity and particle size, and can supply information on ratios of, for example, sand to cement to aggregate.

Macro-specimens can indicate the presence of fibrous materials, and thus suggest the possible presence of asbestos.

With wood products macro-samples show the age of trees, the make up of plywood, the particle size of fibre-boards, and the ratio of the different ingredients involved.

Macro-specimens on natural materials such as skins, fibres, bones, etc. will only supply meaningful information to specialists in these areas.

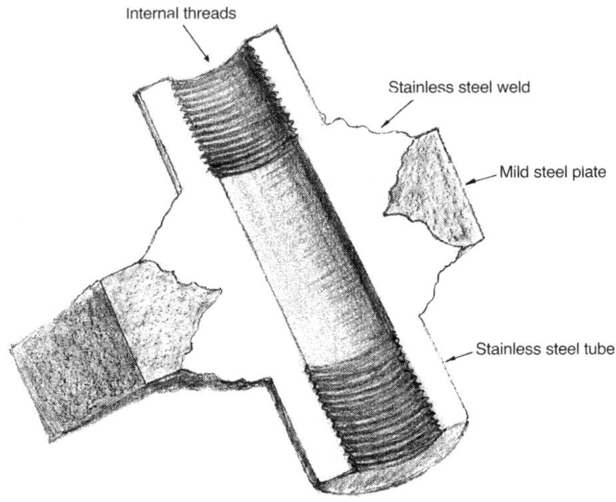

Figure 5.2 Macro-sample of a stainless steel threaded pipe welded with stainless steel to a mild steel plate.

Non-destructive testing (NDT)

6

Non-destructive testing is a common procedure in commissioning and servicing plant or routine production inspection, identifying possible product problems which may then be submitted for failure investigation. It is also used by a failure investigator either as part of the visual examination, or resulting from information obtained from micro-examination.

NDT METHODS

Non-destructive tests include:

- Hardness testing
- Magnetic particle inspection (MPI), also known as magnetic crack testing (MCT), Magnetic crack detection (MCD)
- Dye penetrant crack detection – 'Dye (Di) Pen'
- Ultrasonic (U/S) testing
- Radiography, X-ray analysis
- Techniques for identifying metal differences based on the eddy current principle (for example electromagnetic detection (EMD) or the thermocouple effect
- Condition monitoring

NDT identifies three types of defects – (1) surface, (2) subcutaneous, (3) material differences, as well as an absence of defect as in condition monitoring.

Methods used to identify surface defects are MPI, dye penetrant, and some eddy current methods.

Techniques which cannot find surface defects but will identify subcutaneous problems, that is defects below the surface, are X-ray, radiological, ultrasonic and eddy current.

Eddy current techniques can be used for metal sorting or grading, that is finding differences in material composition or condition. The thermocouple effect is also used for this purpose.

Condition monitoring uses various techniques continuously, or regularly samples and records the results. Once condition monitoring indicates a non-conformance then one of the above three tests will generally be involved.

Where any of the techniques have been used in manufacturing, pre-production, overhaul or routine inspection, then if a defect is identified this could give rise to the component being submitted for further examination. In some ways therefore, NDT is part of the visual examination or first diagnosis routine.

Only experienced qualified inspectors should be called in to perform non-destructive testing, and it is suggested that anyone involving NDT personnel should prompt questions such as 'What is the sensitivity?', 'When was the equipment last calibrated?', etc. The experienced NDT operator will give sensible answers rapidly and efficiently, and the good operator will want to expand on what is seen. The 'cowboy' operator can usually be identified by a few intelligent, probing questions.

As part of the failure investigation either during the visual examination or from other information identified during macro/micro-examination or mechanical testing, the investigator may request non-destructive testing to be carried out on the components in his or her possession, or others in service. The decision whether to carry on cutting and preparing macro/micro-specimens, to have further mechanical tests, or to have a non-destructive test survey for surface or central defects will depend on economic considerations.

If the components are submitted with the information that some defect was found by non-destructive testing, it is possible that the investigator will want to have the test repeated, even if it is only because the surface evidence has been removed or the information submitted is not in sufficient detail. It is therefore important that the investigator appreciates the limitations and also the special benefits of each of the different types of non-destructive tests. We look now at these in turn.

MAGNETIC PARTICLE INSPECTION/MAGNETIC CRACK TESTING/MAGNETIC CRACK DETECTION (MPI/MCT/MCD)

This requires that the material being tested is magnetic and therefore is confined to ferrous metals. Although cobalt and nickel alloys are ferromagnetic they are not normally accepted as having sufficient inherent magnetism to warrant the use of MPI.

Austenitic stainless steels are non-magnetic unless heavily cold worked and therefore cannot be inspected by this technique. Duplex stainless steels may or may not be magnetic: thus they must be checked with a magnet before testing.

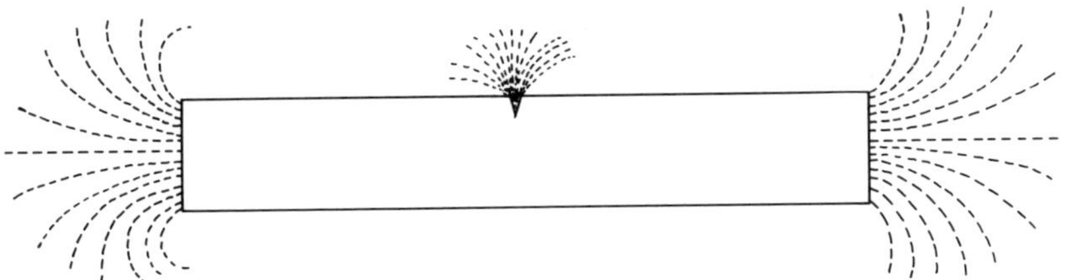

Figure 6.1 Discontinuity in electromagnetic field in ferrous bar caused by surface fault.

Magnetic particle inspection relies on the fact that if there is any break in a magnetic field then poles will appear at this discontinuity. If a bar is magnetized and placed below a thin sheet of paper and iron oxide or iron filings sprinkled on to the paper and the paper vibrated gently, it will be seen that the iron particles move to the poles, that is the ends of the bar. If the surface of the bar portrays a significant scratch or nick then again poles will be produced at this discontinuity, as in Figure 6.1. This is the principle of magnetic particle inspection.

Two points must be appreciated. Firstly that the crack, unless it is considerable, that is is deep and sharp, will not be observable unless the magnetic field is at right angles to the crack. Therefore, with a bar-type component circumferential cracks are identified by placing a horseshoe magnet along the length of the bar. The crack will then be found between the ends of the horseshoe magnet in a circumferential direction. If the cracks are longitudinal then it is necessary to have the field at right angles and this means that the magnet, if it is a permanent magnet, has to be moved round the circumference of the bar with the magnet parallel to the length of the bar. Use can be made of either permanent magnets or more commonly electromagnets, where the yoke is magnetized by a direct current when required. Solenoid coils round a bar with direct current can be used to find longitudinal defects. It is possible to use special test machines but these are seldom available as portable equipment.

Secondly, there will be a size below which a defect cannot be identified, known as the sensitivity limit. In practice if there is a surface scratch, score or crack which is less than approximately 0.05 mm deep it will not normally be identified by magnetic particle inspection. Any scratch, score or crack deeper than this should be found; thus the sensitivity is stated to be 0.05 mm (0.002″). The sensitivity of magnetic particle inspection for surface defects is better than for any other technique, with the possible exception of visual inspection using magnifying equipment.

It must also be appreciated that if there is any surface contamination this will affect the sensitivity. A film of paint 0.1 mm thick will reduce the sensitivity by 0.1 mm. That is, it will require defects greater than that to be identified using MPI. The paint film also hides all surface defects from visual examination.

This appears to be something which is not appreciated by many testers where they almost invariably apply white paint to the surface whether this is required or not. This is an indication of a lack of appreciation of the process being carried out, and should be criticized.

Once a crack, score or other surface defect has been identified it must then be examined by the investigator with a hand-held magnifying glass. Commonly this will be followed by lightly grinding the surface and re-inspecting. If the defect has not disappeared through grinding then it will have a certain minimum depth and an unknown maximum depth. If it has disappeared through grinding then it can be stated to have had a certain maximum depth.

With magnetic crack detection, there is also a potential problem that sudden changes in section, such as very sharp radii or the toe of fillet welds, can concentrate the magnetic flux giving crack-like indications. These will not be found as defects at micro-examination, and will disappear if the radius is redesigned.

It is worthy of note that such spurious defects will not be picked up by the dye penetrant technique, but they could however act as stress raisers to initiate fatigue. Thus for components which operate in a fatigue situation, the identification of a stress raiser as well as actual defects can be useful and should result in redesign. For lower duty components standard dye penetrant testing can be substituted thus overlooking this type of defect.

DYE PENETRANT TEST

This technique should only be chosen when MPI for some reason is not possible or advisable. Magnetic particle inspection will find finer cracks by a factor of at least 2, probably 5, than standard Dye Penetrant; it is also faster, thus more efficient. However, there are available now Dye Penetrant systems which it is claimed have an equal sensitivity to the magnetic test.

Dye penetrant relies on the surface defect being capable of absorbing a coloured liquid, retaining this and thus highlighting the area where the defect existed. The surface to be inspected must be clean and free of excessive roughness. It is sprayed or immersed in a liquid which has certain characteristics, which are firstly that it has a very low surface tension, and secondly can be seen. This is achieved by including a chemical such as methyl salicylate (winter-green) and a vivid dye in the fluid.

The liquid is left in contact with the surface for a minimum of 10 minutes and is then removed by washing which is by far the most critical part of the dye penetrant inspection technique. If washing is excessive then all the dye will be removed and no defects found. If washing is inefficient then the background will be coloured and it will be impossible or very difficult to identify cracks.

It also must be appreciated that fine cracks will hold the liquid better than wide mouthed cracks; therefore, provided the right equipment and type of dye is used, this technique is more efficient for fine cracks which are often more serious than wide mouthed cracks and can be seen by visual examination anyway.

As with magnetic particle inspection the depth of the defect can be identified by grinding or cutting. Depending on a number of factors, including the skill of the operator, the dye penetrant test is popularly estimated to be one-half or one-fifth the sensitivity of MPI.

There are various modifications made to the procedure such as coloured or fluorescent ink, which requires ultra-violet light. These more readily identify the crack but do not improve the 'sensitivity'.

With very limited exceptions dye penetrant crack detecton should not be used where magnetic particle inspection is possible. One exception would be where there are a series of rapid changes in section where the magnetic flux would tend to concentrate and invalidate results.

The failure investigator will commonly require magnetic particle inspection or dye penetrant crack test as part of a cost-effective visual examination to quantify the extent of surface defects. It will be less common that this will be required during the later examination, but macro or micro information may suggest that crack detection is desirable on other pieces of the failed component, or that it is necessary to crack test all similar components in service.

ULTRASONIC INSPECTION

This technique is relatively modern, is basically the same principal as radar, using energy at a frequency above that of sound but below that of radio. The high frequency energy waves are fed to the surface and penetrate to the rear face. There they are reflected and collected by the receiver unit. The result is presented to an oscilloscope with a 'pip' or peak at the entry and exit. If no defect exists between the surfaces there will be a flat line. Any defect of significance will be shown as an intermediate pip on the oscilloscope line. This description is of the simplest possible form but basically all ultrasonic methods work along these lines, see Figure 6.2.

It should be noted that with the main signal and echo at the entry and exit to the sample this technique cannot be used for fine surface defects as these will be obliterated by the entry and exit pips. This technique therefore is for subcutaneous defects and in general is used prior to machining in order to identify whether or not there are material defects which would appear at the surface after machining. Ultrasonics are also used for welding samples where they will readily identify any lack of fusion, porosity or cracking.

As with surface crack detection, it is necessary to have some information regarding the sensitivity, that is the minimum size of defect which can be

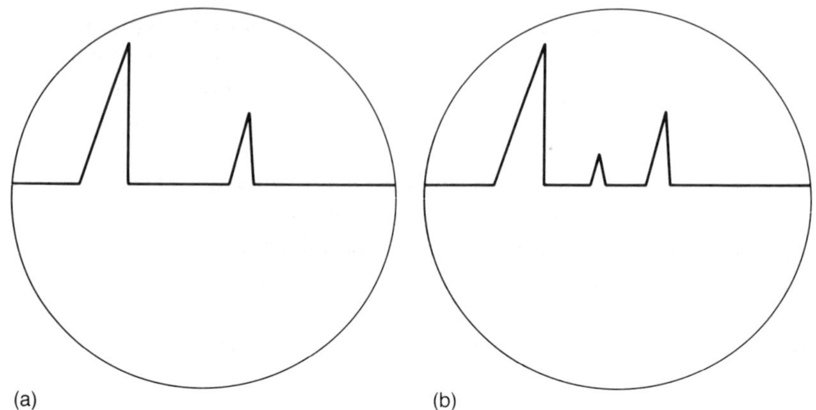

Figure 6.2 Ultrasonic flaw detection in metals. (a) Screen view of reflections from the upper and lower surfaces of sample under test. (b) As (a) with additional reflection or echo from internal defect which gives an indication of position and size of the flaw.

identified. The equipment and techniques for ultrasonic inspection are quite sophisticated and require training and ability of a higher standard than that required for surface crack detection. It is, however, possible to assess whether or not any ultrasonic test operator called in has the necessary skills by firstly describing the equipment and techniques, and secondly determining the sensitivity.

Failure investigation, then, will on occasion be initiated by the findings at ultrasonic inspection. Although the failure investigator may not often make use of this technique as part of a failure investigation his or her report may recommend that components similar to the failure should be subjected to ultrasonic inspection.

RADIOGRAPHY – X-RAY

This makes use of X-rays, which can be compared to light-waves and have the same effect on a photographic plate. That is, where the ray intensity is high there will be a dark spot, where the light is low there will be no effect. There are two sources available, firstly electronic emitters to produce X-rays, and secondly the use of radioactive materials such as cobalt which produce similar type rays.

Basically the priniciple is that the source is located on one side of the specimen with a sensitized film on the other surface. The specified amount of radiation is emitted through the sample, the film is developed and any voids, cracks, etc. will be seen as dark spots or dark lines. Conversely any material denser than the basic material will appear as light spots. It will be appreciated that a square block with a crack through the middle with the radiation applied at right angles to the defect will show a very small defect but if the rays are applied on the same plane as the crack it will show a well-defined dark line.

Thus a book tested from front to back cover will show no defect, but tested from the spine to the front will show each page as a black line. It will thus be seen that there is some skill on the part of the operator, firstly in choosing either a single or multiple shot and then in interpreting the results.

Radiation is highly dangerous and thus requires specialist operators with costly equipment often working unsocial hours. However, this is seldom required during a failure investigation but it may be recommended in the report that components are inspected using this technique either at production, or in some cases in service.

EDDY CURRENT AND THERMAL EFFECTS

All metals are affected by being placed in a direct electric current field. The effect of this 'eddy current' can be shown by connecting two direct current coils to an oscilloscope, one coil to the X-plane, the other to the Y-plane. If the coils and the current are identical this will produce a perfect circle on the oscilloscope tube. If any metal object is then put into one of the coils, an angled straight line will appear instead of the circle. If a metal object identical in all respects to the first is put into the second coil, the perfect circle will reappear. If however, there is any difference in the size, shape, surface finish, hardness, material composition, or metallurgical structure between the two objects, then this will affect the signal to the screen.

Until quite recently this Eddy Current inspection technique had little or no impact outside aircraft engine manufacture, or other industries which have very tight tolerances of dimensions, and control of material differences. The variations to the signal resulting from minute differences in shape and size made it impossible to identify any significant difference in surface, hardness, or metallurgical structure. With modern electronic magic power, it is now possible to 'tune out' any parameter, or parameters, which are not relevant to the inspection. There are a limited number of companies supplying this sophisticated equipment, now known as EMD (electromagnetic detection). There is no doubt that in the correct hands this technique can rapidly inspect for surface finish and cracks, material differences, hardness variations and dimensional deviations.

It will be seldom that the technique can be applied to a failure investigation, but there could be occasions where the investigator might recommend the use of this equipment for components in service. As with ultrasonic and other non-destructive tests it is imperative that those supplying the service can and do explain the limitations, sensitivity and other aspects of the equipment.

Electromagnetic detection (EMD) in the correct hands can be a useful, highly economic tool for finding a variety of defects.

All the above non-destructive tests are used under various conditions and there is evidence that the standard applied can vary dramatically.

The skilled investigator should be aware of these variations, and be prepared to oversee or supervise any non-destructive testing which is recommended, to ensure this is carried out to the necessary standard.

Unlike the eddy current effect the thermal effect control relies on the fact that when two different metals are connected together and heated, a voltage potential will be generated. This will vary for different metals and always increases with temperature, and is the principle used to measure temperature by thermocouples.

There are now available instruments with a heated probe at a fixed temperature, which when brought into contact with another metal a voltage is generated. This is used to sort components of similar shape and size which could be made from different materials. All identical pieces will generate the same voltage. Any with dissimilar metal will give a different signal.

This is unlikely to be used in failure investigation.

CONDITION MONITORING

This is a sophisticated form of non-destructive testing which is a modern form of process control or in-line maintenance. It provides a continuous assessment of specific parameters on high duty equipment. It takes various forms, not all of which will be involved at all times. These are:

- Noise measurement
- Vibration measurement
- Chemical analysis
- Electrical or resistance measurements

In some instances equipment can be continuously monitored with signals given when any parameter moves outside laid down limits.

More generally there is equipment fitted which is used to monitor the parameters at intervals, or samples are taken at specific intervals, or after certain actions have taken place.

Many of these measurements are involved with environmental or public health, and seldom result in a failure investigation. Those which measure noise, vibration, chemical analysis of moving machinery, or electrical resistance, voltage or current flow in cathodic, or impressed current corrosion protection, quite commonly identify failure at an early stage. The failure in these instances will be 'deviation from performance' and very seldom will the investigator be presented with fractured or broken components.

Noise

Where noise is measured this will be to monitor the total noise level, but also in some cases will analyse the level of noise at various frequencies.

This has been found to be more sensitive than vibration. It can result in many investigations which show no problem exists, but this is more economical than investigating broken bits. The equipment is specialist and requires considerable planning and experience in interpreting the results.

There is a case history where a ball bearing was investigated for noise where stress-corrosion cracking of the track was found. If this had progressed to cause either vibration or failure, it is doubtful if this would have been identified as the evidence would have been destroyed. The report encouraged further investigation which ultimately identified contaminated oil which could have resulted in seizure.

The wheel tapper of the railway age used to find cracked wheels by the different sounds made when the wheel was struck with a hammer.

Vibration

This is probably the most common technique used in condition monitoring. Sensors are fitted at critical points and the vibration recorded at all levels of use of the equipment. Any deviation from the standard vibration is investigated. This will identify, for example, a fatigue crack propagating before fracture occurs.

Ball, roller, and plain bearings will all start to vibrate long before seizure or break up takes place, allowing a routine change rather than stoppage. It can also give the failure investigator a much better chance of identifying the cause of the problem.

Chemical analysis

The routine analysis of oil is an early example of condition monitoring. With this a sample of oil is analysed before use for specific elements, using modern mass spectrometry techniques. Samples are then analysed at intervals of operation and compared with the initial sample. This can show a change in characteristics, such as viscosity, but also the presence of metals. These can pinpoint quite accurately where any excessive wear is taking place and the necessary steps to take.

This method of failure is routine on items such as helicopter gear box oils and for heavy duty engines of all types. Results of the analysis may be used for routine maintenance in lieu of either mileage or operational time checks. If a failure has occurred the result of the analysis can supply valuable information to the investigator regarding the mode and possible reason for the failure.

Electrical

The majority of such measurements will be for control purposes, including resistance and temperature measurements for metallurgical and chemical processes, and potential measurement for corrosion protection. They can be of considerable assistance to the metallurgist or chemist investigating a failure which could be the result of poor control at heat treatment or corrosion protection processes.

Monitoring devices include strain gauges, where variation in stress levels can be identified on critical components, and 'rub indicators' which can be used to show that two metal components are making contact when there should be clearance.

Mechanisms of failure

7

There are six major failure mechanisms to consider in all investigations of mechanical failure.

1. **Tensile** This covers fatigue, bending, brittle and ductile fracture and is by far the most common mechanism. The investigator will be able to identify by visual examination the different types of tensile failure, describe the characteristics and indicate the next stage of investigation to consider.
2. **Compression** This seldom results in fracture but can cause distortion. Compression, that is failure due to pushing forces can also result in a tensile or bending force being applied, resulting in tensile-type failure.
3. **Shear** This includes the guillotine type action and torque – a combination of tensile stress and compression. The fracture face will identify the detail. This failure mechanism, together with the previous two, are the mechanisms to be identified when fracture or distortion exists.
4. **Corrosion/contamination/degradation** Corrosion can result in components being removed from service as they are no longer fit for the purpose. Corrosion itself can initiate fatigue, or be the stress raiser for brittle fracture. Contamination can affect products in contact with the surface, causing corrosion of metals, and surface attack on non-metals. It can result in abrasion when the contaminating material is hard, or the corrosion products are abrasive. Degradation is the result of chemical changes influenced by heat, light or temperature over time.
5. **Abrasion/wear/erosion** This is where excessive movement, contamination, noise, heat, etc. are identified and may result in fracture or seizure or poor performance and thus requires investigation.
6. **Thermal – heating or cooling** Where components are subjected to a temperature above or below that specified, the change to the material structure and dimensions can dramatically alter the performance properties. Increase in temperature above a certain figure will normally reduce the tensile properties. Reducing the temperature below a critical figure may increase the tensile stress but can reduce ductility and increase brittleness

dramatically. The effect of heating and cooling on plastic/polymer materials is much more significant than with metals.

It should be possible for the skilled experienced investigator to identify one or more of the above mechanisms during visual examination of the failure before further evaluation is commenced. It must be emphasized that it is the quality of the visual examination and the ability of the investigator to see and recognize the minute detail that are of paramount importance in failure investigation. If the critical factor is not recognized, it will be overlooked and not be investigated.

The mechanism of failure is the 'How'. The next stage is 'Why'. The remainder of this chapter will describe and illustrate the different effects seen, discuss the variations, and give options for further examination and investigation.

TENSILE FAILURE

This covers ductile and brittle fracture, fatigue and bending.

Ductile tensile failure

This will have well-defined characteristics generally in the form of a reduction in area at the fracture, sometimes known as 'necking', and there may be the appearance of 'ears' associated with the fracture. These features indicate that the material has been taken above its yield strength and has stretched to a measurable degree beyond the yield point to give the ultimate tensile fracture. This will indicate that the material has a good degree of ductility and has been subjected to a tensile load probably applied in a reasonably slow manner and without impact.

The investigator will require to identify whether or not this is the correct material for the purpose intended or is in the correct condition. It will be necessary to find the original material specification for the part or failing this what would be expected for a component of the type which failed. All that might be necessary then would be hardness tests and micro-examination to indicate the ultimate tensile strength of the material. Hardness tests in conjunction with a micro will supply the practical metallurgist with sufficient information to inform his client whether or not the correct material has been used.

For example, if the material should have been an alloy steel with a tensile strength in the region of 900 N/mm^2 (60 tons/in^2) and the micro-examination shows a plain carbon steel with a hardness relating to a tensile strength of 500 N/mm^2 (35 tons/in^2), then the wrong material has been used and no further action is necessary. A report can then be produced.

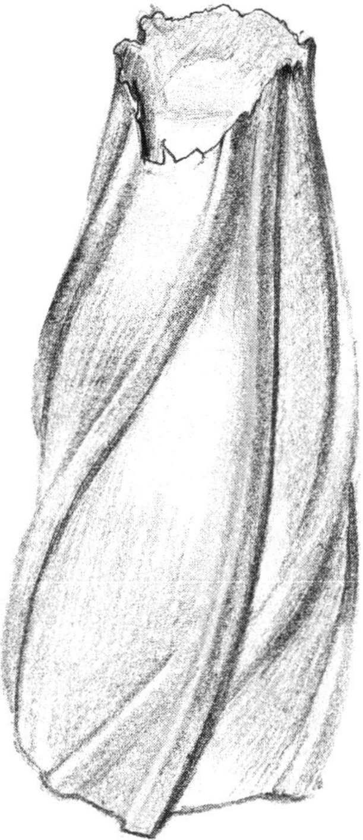

Figure 7.1 Ductile form of tensile failure in steel showing typical reduction in cross-section (necking), and a visible crater with 'ears' at the outer diameter.

Only steels can have the hardness value related accurately in this way to tensile strength, other metals having a much less accurate relationship. Non-metals will exhibit the same visual characteristics, but may have no relationship between hardness and strength to serve as an indicator of failure.

The ductile failure illustrated in Figure 7.1 was in a 25 mm steel reinforcement bar for concrete.

Another example of a ductile failure is the investigation of the sinking of a hopper barge. This was used to collect the dredgings from an estuary. The system used was that the barge was held alongside the dredger with three clips on the gunwale (gunnel). The dredge had 'U' clips, with the hopper inverted 'M' clips designed to slide into each other as the barge was brought alongside.

Failure of all three clips occurred, allowing the barge to float free causing considerable damage before sinking. Investigation showed that the clips on the barge had been bent with the visual evidence on both clips and dredger indicating that instead of sliding into each other, the edge of the hopper clips had been sitting on top of the dredger clips.

As the hopper was filled, instead of sinking lower in the water, the top of the clips experienced loading, firstly bending, then fracturing two of the three clips. The fractures were ductile adjacent to the weld of the clip to the gunnel. Micro-examination showed an excellent weld, and confirmed that the failure was ductile with evidence of cold work but no material defect. Hardness tests showed the material was of a tensile strength of 500 N/mm^2 approximately, suitable for this application.

This therefore is an example of a correctly manufactured and welded component overloaded by misuse and failing in a ductile manner due to tensile forces.

Brittle tensile failure

This has a flat fracture face as in Figure 7.2. There will be no indication of any reduction in area associated with the fracture. This will tell the investigator that this component at one instant was operating in a satisfactory manner, and in the next instant had failed. Once the fracture face has been cleaned it can be examined in detail. This may show fatigue, which is discussed below.

Brittle fracture can exist without fatigue, however. The surface should be examined for grain size with large grains being an indication of a brittle

Figure 7.2 Brittle failures from the same bar as in Figure 7.1. This fracture has a flat face with no reduction in cross-section. The fracture shows lines or striations, indicating the point of initiation. The cause was the result of a stray electric weld.

material. Fine grains may not indicate ductile material but large grains will always have a tendency to be more brittle.

The fracture face must be examined for any indications of directional fracture. This will be in the form of faint lines which require some skill and experience to identify but will indicate that a bending moment has been applied as well as giving the direction of this force. If this is noted then the fracture faces can be placed together to see if any bending has taken place or whether the fracture has occurred without bending. The investigator will be able to see the initiation point on the surface where fracture started, and this point must be examined for any stress raiser such as corrosion, damage, weld, etc.

Micro-examination will be necessary to find grain size, excessive oxide or slag, or a brittle metallurgical structure. The investigator would also probably require a hardness test, impact, and perhaps a tensile test.

The example shown in Figure 7.2 is of brittle failure of a failed steel reinforced bar for concrete. This could be seen to have a weld spatter, or a stray arc at the initiation surface of the failure.

Other brittle failures are the result of faulty heat treatment, generally with some form of stress raiser initiating fracture.

Where a brittle fracture has been identified and no metallurgical, mechanical or analytical reason can be found, then hydrogen embrittlement must be considered. Hydrogen embrittlement results in a brittle tensile fracture mechanism, which is difficult to verify. The experienced metallurgist, presented with a brittle fracture and certain information, would possibly conclude that hydrogen embrittlement was the cause of the failure. It is not, however, possible to present evidence by any known technique at present to prove the embrittlement was caused by hydrogen.

It is now known that the problem is confined to high tensile (invariably steel) materials which have been involved with atomic (nascent) hydrogen, that is H. (Note that hydrogen sold in bottles – pressure vessels – is molecular hydrogen, H_2. The substance which initiates embrittlement in steel is atomic – nascent – hydrogen, H.)

The theory of hydrogen embrittlement as presented, and accepted by the author, is that the atomic hydrogen enters and fills all the interstices of the metal. It then converts to molecular hydrogen (H_2). This appears to have a larger volume for the same mass and thus exerts an internal stress. Low strength metals have sufficient elasticity/ductility to yield, whereas high tensile steels do not, and fracture when any appreciable stress is applied. This can be likened to a stoppered bottle filled with gas, which then increases in volume. A glass bottle will fracture, while a rubber or plastic bottle will expand.

This simple theory appears to fit all the facts known regarding hydrogen embrittlement, since it only affects high tensile materials which have been involved with nascent hydrogen. It does not affect ductile metals, even when nascent hydrogen exists, and can be eliminated by heating components immediately after they have been subjected to atomic hydrogen.

Thus the investigator presented with a failed hardened component which has been recovered by chrome plating or perhaps with broken self-tapping zinc plated screws is safe to assume that hydrogen embrittlement is the probable cause.

There is less reason to attribute welding failures to hydrogen embrittlement. Many of these are the result of heat-affected zone hardening, which subsequently crack or fail because of excessive hardness. Oxyacetylene welding and brazing never present problems which could be blamed on hydrogen, but can still have problems with cracking in the heat-affected zone.

A weld failure investigation which reported hydrogen embrittlement as the possible cause of failure and also identified high heat-affected zone hardness above 300 DPN should be suspect, as the high hardness could result in cracking without hydrogen.

Considerable sums of money have been, and are being, spent in 'controlling' hydrogen embrittlement associated with welding. These are so-called 'low hydrogen' welding rods supplied at some expense in vacuum packs for welding mild and low alloy steels which would not be susceptible to embrittlement.

Brittle failures with non-metals such as plastics/polymers will show similar characteristics to metals, while concrete, bricks, etc. will always show brittle characteristics. Glass at normal temperatures will fail in a brittle manner, but will distort at high temperatures. Microscopic examination might be useful in some cases, but cannot contribute to the same degree as with metals. Hydrogen embrittlement will never be a reason for non-metal brittle failure.

Fatigue failure

Fatigue is due to the application of a cyclic tensile load. The component may be within its design yield strength, but because of the presence of a stress raiser, a small portion of the surface will be taken above the ultimate tensile strength of the material, resulting in a crack. The crack will propagate each time the surface is taken above the ultimate tensile strength at that point. This will be below the proof strength of the component in most cases.

Fatigue fracture can be recognized by striations, or 'beach marks'. These will indicate the initiation point or surface fault.

If, for example, the crack starts at a sharp radius or change of section as in Figure 7.3, then this will tell the investigator exactly at which point the load was applied, and the striations will propagate from this point. If the striations follow round the circumference, and propagate towards the centre, this will indicate 'concentric fatigue'. This is the result of either a very sharp radius or an overload, perhaps from a bent shaft.

The striations have the great advantage in that they help to identify the starting point, and allow the investigator to concentrate in this area. Potential

Figure 7.3 Fatigue fracture at the point of a sharp radius illustrating the smooth fracture face with striations from the initiation point.

problems such as lack of radius, corrosion pitting or prior cracking for some reason will concentrate the metallurgical investigation even further. Micro-examination at or through the initiation point may then show stress corrosion, slag, local hard spots, etc.

Where the fatigue crack propagates at an angle, up to 45° to the surface, this will tell the investigator that a torque load has been applied, as in Figure 7.4.

Where the fatigue striations follow the circumference of the component, this is 'concentric' fatigue, as shown in Figure 7.5. This generally indicates a high bending load on a rotating component, very often at the point of a sharp radius, or an overload, perhaps from a bent shaft.

Fatigue cracking provides an indication of the safety factor involved with the component on the assumption that the component was working under normal loading conditions. A fatigue crack which propagated 5–10% of the cross-sectional area before fracturing would indicate a very low safety factor, in the region of 10% which is inadequate. A fatigue crack which propagated 20% of the cross-sectional area before fracturing would indicate a 20% safety factor. This is the minimum safety factor normally accepted with aircraft-type components where weight must be saved and components work to a higher standard than normal.

Where a fatigue crack propagates approximately 75% of the cross-section prior to fracture, this would indicate a normal safety factor for engineering

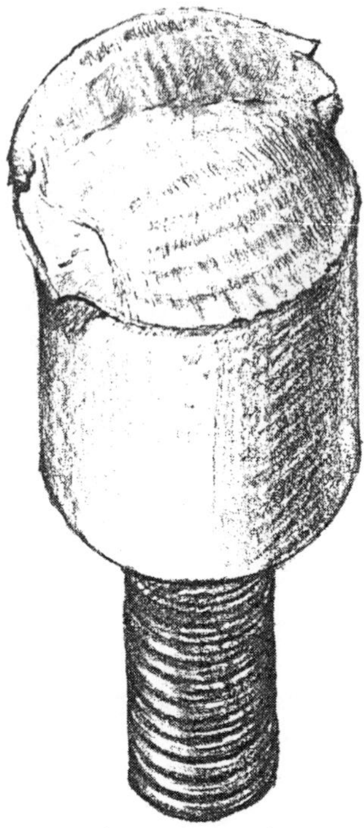

Figure 7.4 Fatigue fracture from a sharp radius, with the crack propagating across the section at approximately 45° indicating the presence of torque loading.

components. A fatigue crack over 90% or more of the cross-sectional area indicates a very high safety factor and once the client is informed that the component was operating with the cross-section almost 100% cracked, advice can be given that the load to cause fracture is very low. The component can perhaps be redesigned to save weight provided the reason for the fatigue initiation is clearly identified and eliminated. Figure 7.6 illustrates various degrees of crack propagation in fractured components.

This mechanism was termed simply 'fatigue' as the early investigators were aware that the component had been in service for some time and that generally no metallurgical reason could be found as the cause of failure. Thus old-age or the metal becoming 'tired' appeared to be the reason. It was not until the Comet aircraft disasters were investigated that the importance of stress raisers was fully appreciated. With a smooth surface and a gradual change of section, any tensile load applied to the surface will be evenly distributed across the surface. Where, however, there is a sharp radius, pitting, damage marks, cracking, brittle metal, soft metal, etc., then the same load, instead of being spread over a surface, can concentrate at one point or area, and thus the unit

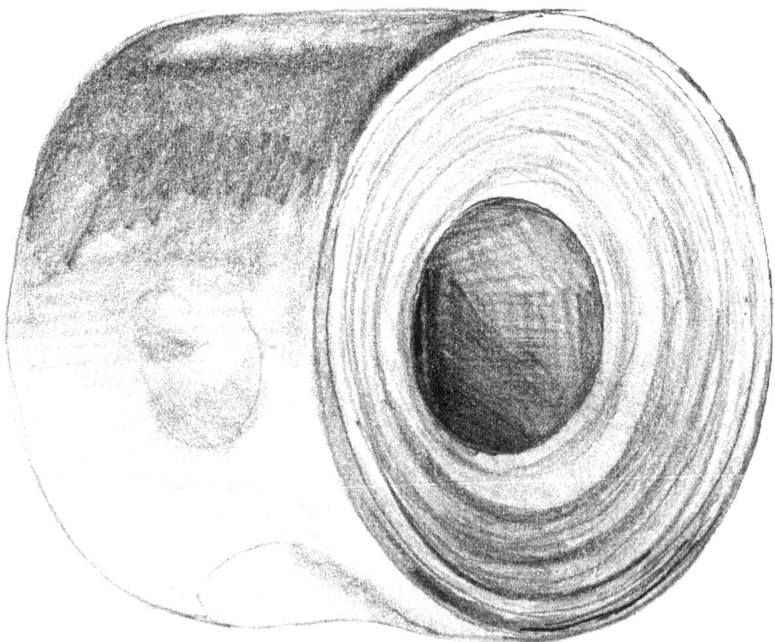

Figure 7.5 Striations visible in a fracture surface which are in the form of concentric circles. This usually indicates a high bending load on a rotating component.

stress applied will be much higher and can exceed the tensile strength at that point or surface.

The fatigue strength can thus be dramatically improved by removing the stress raiser, and it is the responsibility of the failure investigator to firstly identify fatigue as the mechanism, and then to find any stress raiser and advise how this can be eliminated.

Where fatigue failure exists and no stress raiser has been found, on a correctly made component then the investigator must recommend an increase in tensile strength of the material, or in some cases 'shot peening' may be suggested. Shot peening was developed for certain components on piston aeroengines, notably valve springs, where the material strength was at the maximum possible, the surface finish was free of stress raisers, and there was no room to increase the spring wire diameter. Tradition has it that the original theory of shot peening was that this resulted in so many surface stress raisers that the stress did not know where to concentrate! It is now known that shot peening produces a compressive stress on the component surface. As fatigue is a tensile problem, in order to cause failure on a correctly shot-peened surface, it is necessary first to overcome the compressive stress present on the surface.

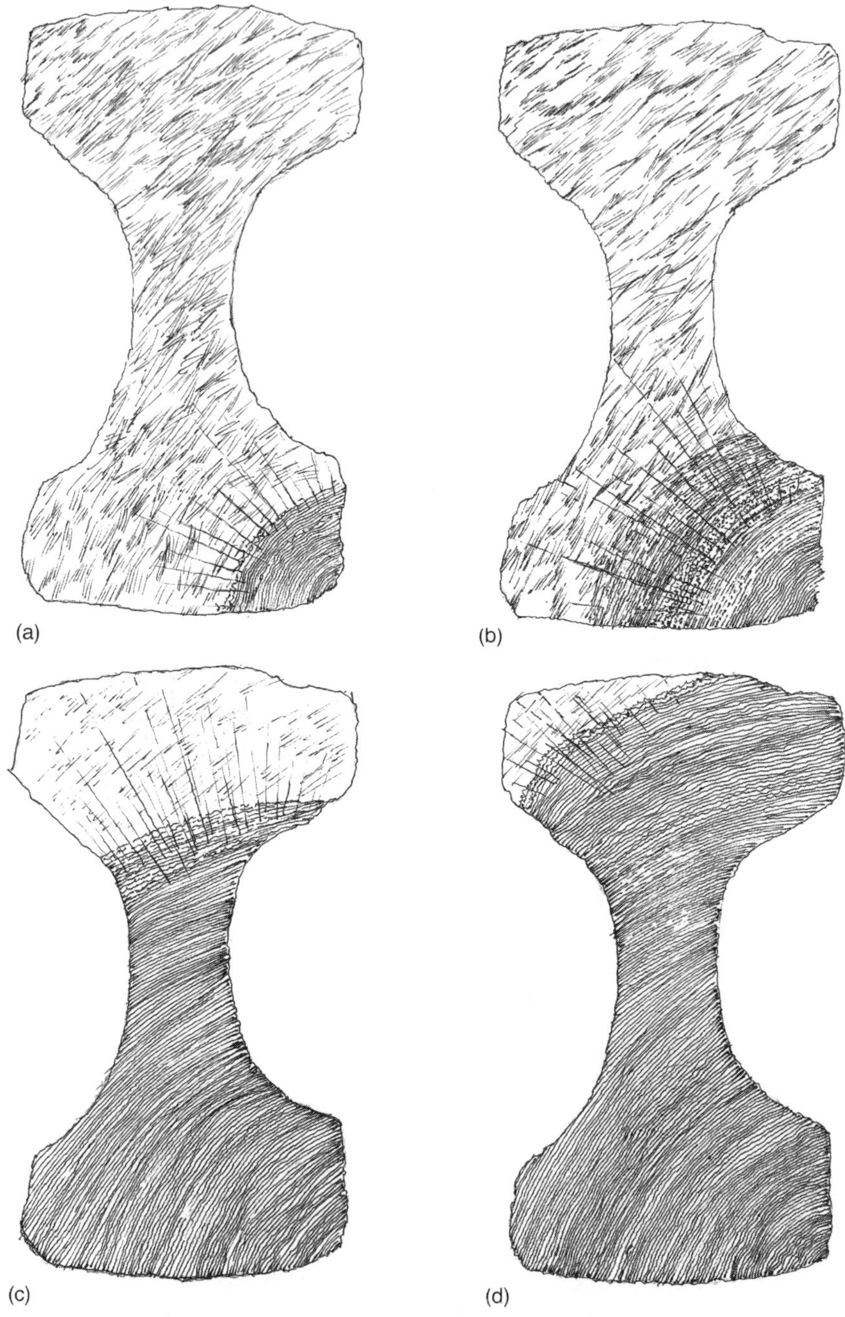
(a)
(b)
(c)
(d)

Thus shot peening increases the fatigue strength of a component. It is applied by bombarding the component surface with hardened steel shot under controlled conditions. The shot must be round so it does not cut, and the surface to be peened must be free of all stress raisers.

There is a case history given where the importance of eliminating all stress raisers prior to shot peening was not fully appreciated. The failure investigator recommending shot peening must ensure that the client understands that although the finished shot-peened surface can be quite rough, the surface prior to peening must be of a high standard, and must explain in detail the reason.

Only plastic/polymer materials, including rubbers, of the non-metals will show visual fatigue characteristics, but there is no doubt that concrete and probably wood fail by the fatigue mechanism described although the fracture surface does not always clearly show the typical fatigue striations.

It is doubtful if there is the same relationship between fatigue and tensile strength with non-metals as is known to exist with metal fatigue. Concrete, bricks, etc. do, however, fail by fatigue, with very often the surface spalling (coming away in large chips) when subjected to tensile loading. This can be the result of suction as wheeled traffic passes over any surface imperfection.

Bend failure

Bending will not of itself result in fracture. The failure mechanism involved with bending will be the ductile or brittle fracture described above. When a component bends beyond a certain degree, it will no longer be fit for its purpose. The bend will be identifiable by visual examination. The investigator can estimate the direction of the load and then look for cracks, evidence of fretting, corrosion or other effects which could have weakened the section. The fact that it bent would indicate a degree of ductility.

Figure 7.6 Cross-section of car connecting rod showing fatigue cracking at a corner from a slag stringer. (a) Fatigue cracking propagating across approximately 10% of the cross-section before fracturing: failure thus occurred at only 10% of the design load. (b) Fatigue cracking at 25% before fracture: this would be typical of aircraft-type components where the safety factor must be low for weight reasons, but with high quality control for safety reasons. (c) Fatigue cracking approximately 75% before fracture, typical of a normal engineering component failure. (d) Fatigue cracking approximately 95% before fracture, indicating an unnecessarily high safety factor.

Figures 7.7 and 7.8 show a bent shackle pin.

Hardness tests and micro-examination would tell the investigator whether the correct material had been overloaded or the wrong material used.

Some non-metals such as glass and concrete will not bend. Others such as rubber and man-made fibres can be tied in knots without causing a problem.

Many plastics/polymers have excellent ductility properties with significant stiffness, and when used as composites have a range of properties now competing with metals. Bending can, however, result in failure of the bond or component. This will be identified at visual examination and may be in

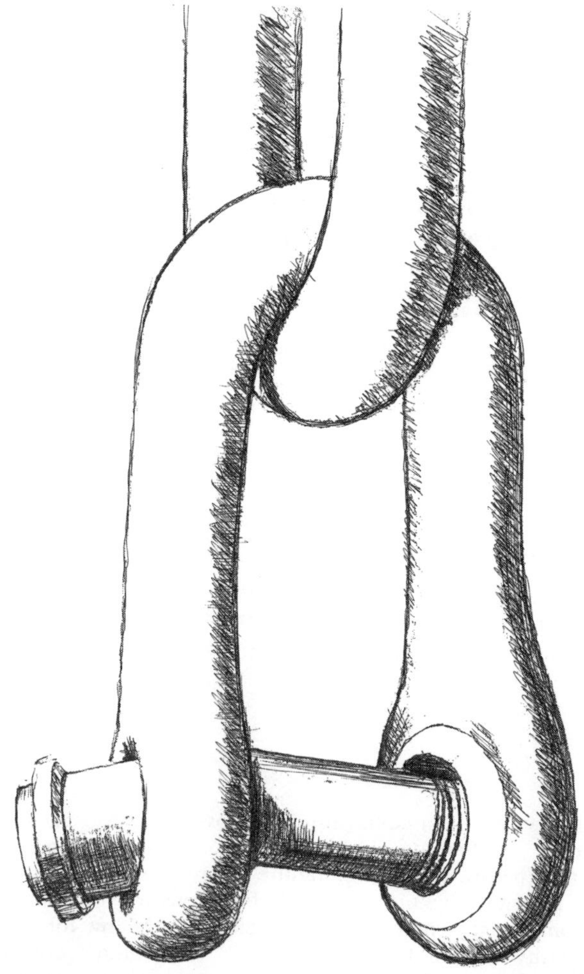

Figure 7.7 A shackle assembly which had been overloaded, bending the pin. This resulted in the shackle jamming.

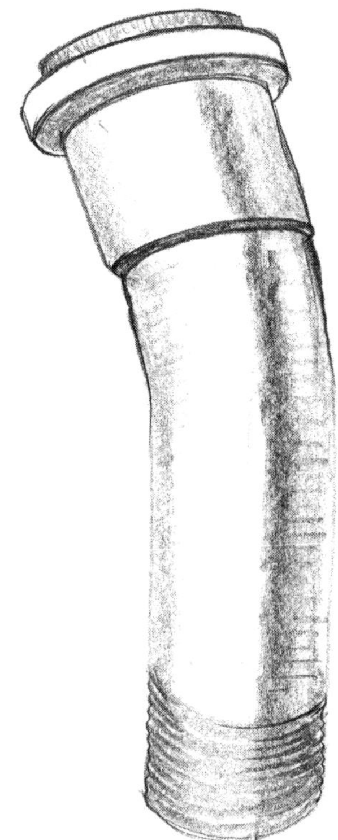

Figure 7.8 Showing the bend on the pin removed from the shackle in Figure 7.7.

fact a fatigue failure. It would be unusual, however, for a non-metal component to be investigated because of a permanent bend.

With non-metals there is the dramatic comparison of rubber, wool, and leather which can be tied in knots without causing any problem and glass, brick and stone which will crack or fracture when any bending moment is applied. The plastic/polymer materials vary over the range, from rubber-like to as brittle as glass.

Glass, stone and brick will always be designed to operate without bending, and can be expected to fail quite dramatically if subjected to bending.

COMPRESSION FAILURE

This is seldom the reason for fracture. Compression causes the component to be pushed not pulled. The yield strength under compression is exactly the same as tensile. When a component is subjected to tensile loading and the yield strength is exceeded, there will be a reduction in the cross-sectional area, thus the unit load will increase and fracture will take place if the same load is continued to be applied. With compression, when the yield strength is exceeded the cross section is increased, thus for fracture to occur the load must be increased further. Failure by compression will thus seldom result in fracture.

Exceptions to this could be components with threads or splines, where a pushing load can result in shear or tensile loading of the threads or splines. This will be identified at the visual examination if the direction of loading is known.

Components sometimes fail by compression, however, because of distortion increasing the diameter. Damage can also occur with local high pressure being applied which will swage the surface locally. The investigator will need to check the tensile and yield strengths of the material in this case. Figure 7.9 illustrates this condition.

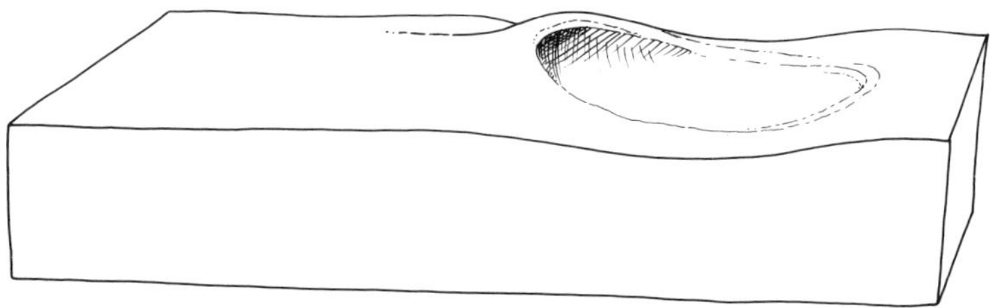

Figure 7.9 A surface which has been loaded in compression, resulting in local metal movement similar to swaging.

With non-metals such as rubber and some soft plastics/polymers compression results in distortion which disappears when the load is removed. Concrete, stone and brick are designed to work in compression, but failure can occur through spalling of the surface.

SHEAR AND TORQUE FAILURE

Shear failure

This is quite rare as a failure mechanism in the form of pure shear, which requires the component to be held as a guillotine or punch. Shear is a combination of compression and tensile stress. For pure shear there must be no bending. A shear fracture will be at right angles to the diameter or section and will be flat with no evidence of any directional properties. There may be evidence of small 'ears' where final parting takes place. The shear fracture face is smooth. It is difficult to describe or illustrate but the experienced investigator will recognize shear when it is present.

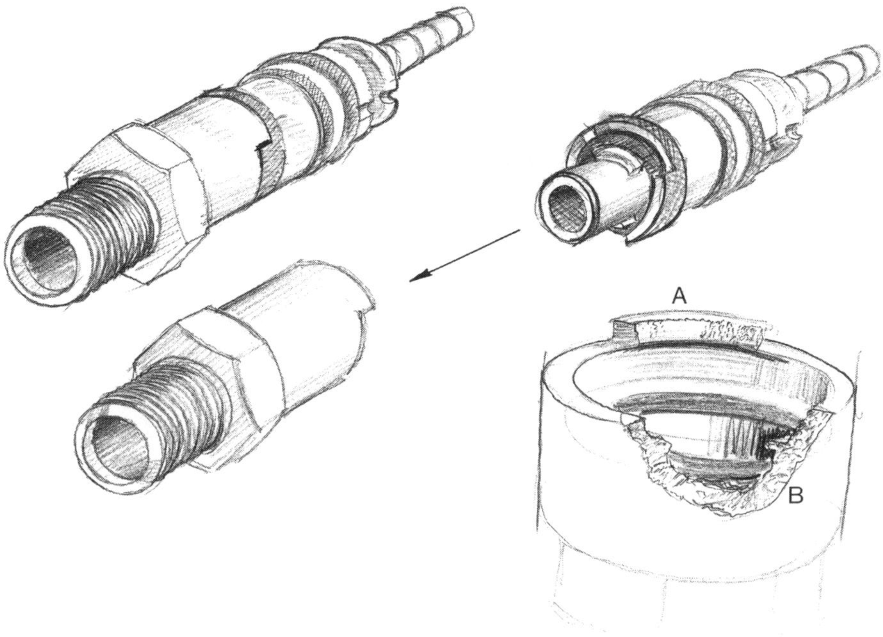

Figure 7.10 Shear failure in a quick-fit coupling. This was subject to excessive side load resulting in pure shear of lug A. Lug B then failed by bending/fracture, releasing the coupling.

It may be found that some bending has taken place prior to shear, as the fit is generally less than perfect. Where bending is slight, then final fracture can be by shear. When significant bending exists, failure will involve tensile failure on the outer surface followed by final shear failure. All these actions must be looked for and identified. Some shear failures exhibit fracture of a fibrous nature indicating that tensile failure is involved on that section of the failure.

Pure shear is a relatively rare mechanism of failure. The shear strength of a metal is lower than its tensile strength, but because of the vagaries of the mechanism when shear is involved it can be difficult to quantify.

It is worth remembering that probably the most common lay-person's description made regarding broken pieces, is that they have 'sheared' or 'sheared off'. To a metallurgist, shear failure is quite distinct from tensile or compression failure.

Shear failure of a quick-fit coupling is shown in Figure 7.10.

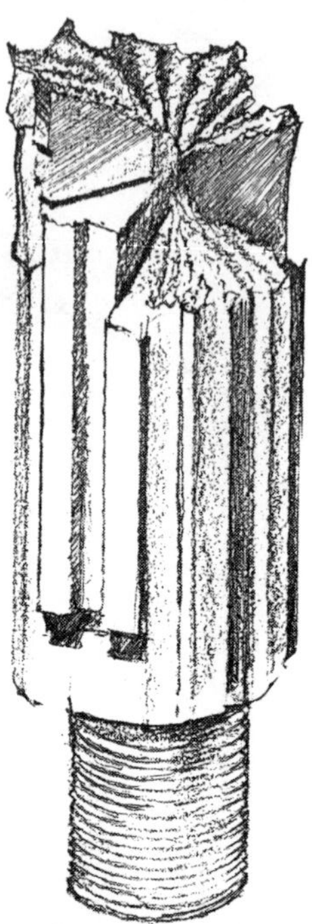

Figure 7.11 This shows impact-type torque fracture with a 45° angle, and no evidence of ductility but does show radial cracking.

Torque failure

Torque is a form of shear where twisting has been applied. Where pure torque is involved, that is where one portion of the component has been static and the other has been subjected to twist, there will be a 45° angle tensile-type failure. Torque loading exists with other forms of failure, for example, ductile tensile where there has been additional torque applied. This will have a distinct twist and the investigator should identify this combination. With fatigue shear this will propagate at an angle which in general will be 45°.

Where surface cracking is seen at an angle to the diameter it can be assumed that torque loading has been present. The investigator can identify from the visual examination the direction of the twisting force to cause the cracking.

Impact torque where the component has a smooth surface will result in bursting-type cracks parallel with the fracture. Where there are surface stress raisers, for example splines, the impact torque will cause radial cracks from

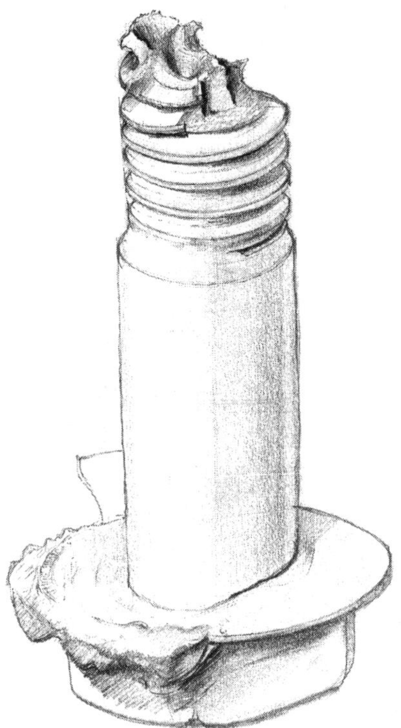

Figure 7.12 Ductile torque failure. This shows the ductile 'ears' with 'necking' and a 45° angle fracture.

the surface with final failure at 45° angle. If however, the impact torque load is high, the radial cracks will meet at the centre.

Various views of torque failure are given in Figures 7.11 to 7.14.

With non-metals shear is comparatively rare, but torsion failures can be common with plastics/polymers and wood.

Glass, stone, brick and concrete seldom fail through shear or torque, simply because they are not designed to experience such conditions: none of these materials can be expected to withstand even quite low torque loading. Glass would fracture, brick, stone and concrete would have spalling failure of the surface.

These materials are, however, commonly subjected to shear load in building structures, with little history of failure.

Figure 7.13 Fatigue initiation with torque connotations, propagated by ductile torque.

Figure 7.14 Fatigue/torque propagation from surface corrosion.

CORROSION/CONTAMINATION/DEGRADATION

There is little doubt that corrosion results in very heavy costs to industry and commerce, and that corrosion consultancy is a source of income to many chemists and metallurgists. The following ode could be the metallurgists/chemists hymn to corrosion!

Rust's a Must

Mighty ships upon the ocean
Suffer from severe corrosion;
Even those that stay at dock side
Are rapidly becoming oxide.
Alas, that piling in the sea
Is mostly Fe_2O_3
And where the ocean meets the shore
You'll find there's Fe_3O_4.
'Cause when the wind is salt and gusty
Things are getting awful rusty.

We can measure it, we can test it;
We can halt it or arrest it;
We can gather it and weigh it;
We can coat it, we can spray it;
We can examine and dissect it;
We can cathodically protect it.
We can pick it up and drop it
But heaven knows, we'll never stop it.
So here's to rust: no doubt about it,
Most of us would starve without it!

Corrosion is a chemical attack on the surface of metals and is less often associated with non-metals such as plastics/polymers, wood or concrete. The term ‘corrosion’, however, is on occasion applied to the chemical attack of concrete, plastic or wood when caused by contamination, where extraneous matter touches or adheres to a surface.

Degradation is the term which is applied when the material of the component, rather than the surface, is impaired. Plastic/polymer materials can also have their properties significantly altered by heat, and to a lesser extent light. The resulting performance will almost invariably be impaired or degraded.

When corroded/contaminated components are removed from service it will be the result of visual examination, as this is always evident as a surface effect.

It is common that corrosion initiates other mechanisms of failure such as fatigue or tensile failure because the corrosion has reduced the cross-section of the component, or acted as a stress raiser.

There are three basic types of metal corrosion:

- Chemical
- Galvanic
- Electrolytic

It is sometimes possible for the visual examination to identify certain characteristics which will point the investigator in a certain direction regarding which of the three is the principal mechanism of failure, but it is often impossible from visual examination alone to identify positively the reason for the failure. This would require micro-examination, chemical analysis and probably some further investigation of the local atmosphere.

There is a large vocabulary related to adjectives on types of corrosion. These include sour, sweet, oxygen, stress, crevice, pitting, bimetal, intergranular. They invariably relate to chemical, galvanic or electrolytic corrosion processes.

Before looking at these three basic types of corrosion it should be remembered that surface contamination which has not resulted in a failure at source might contaminate a product in contact with the equipment, and be a reason for removal from service.

Chemical corrosion

This includes atmospheric attack, over-heating resulting in oxidation, and attack because of impurities physically in contact with the surface. Chemical attack will normally be relatively homogeneous but may be in the form of pitting, or confined to contact surfaces.

With all corrosion investigations, it is essential that handling is carefully carried out to prevent cross-contamination. Chemical corrosion will be the result of surface contamination and if this is removed by cleaning or confused by cross-contamination, then the investigation becomes difficult or impossible regarding pinpointing the initial cause.

Visual examination will have identified pitting which may be discoloured, or flaking where the flakes of material are readily removed. Scrapings or swabs should be taken for further investigation.

The combination of corrosion plus stress can give the typical cracking shown in Figure 7.15.

Figure 7.15 Stress-corrosion cracking. This is an austenitic stainless sheet with a tensile load applied at the centre and chloride contamination. This shows the typical bifurcating cracking pattern. The corrosion in this instance was identified as chemical, but the corrosion could also have been galvanic or electrolytic.

Chemical corrosion will result from attack by an aggressive environment or liquid with the temperature and concentration being critical. Almost invariably the higher the concentration and temperature, the more the attack. Certain chemicals will be more aggressive with some metals than others and the pH, that is alkaline or acid concentration, can be critical. The pH range is: 1 – very acid, 7 – neutral, 14 – very alkaline.

There is now considerable information on the conditions which cause chemical corrosion on each metal. Some metals corrode under very lightly contaminated conditions. This corrosion can be accelerated by any galvanic or electrolytic condition which might be present.

With non-metals surface contamination can result in attack sometimes called corrosion. Sulphur and chloride compounds can attack brick and concrete. (Carbonate can disfigure, sometimes with attack, but generally only producing visual defects.)

Plastics/polymers are attacked by many organic compounds which would not affect metals, and some in fact prove to be useful cleaning agents. They are in general, however, less visually affected by the normal atmosphere. There can, however, be chemical changes – polymerization – accentuated by heat and sunlight which radically affect the properties of plastics/polymers. This cannot really be classed as corrosion/contamination but degrades the properties of some materials.

Galvanic corrosion

This is where two different metals are in electrical contact with a conductive fluid known as an electrolyte. This results in a cell or simple electrochemical battery, the power for which is derived from the electropotential difference between dissimilar metals.

Table 7.1 Electrochemical series

Metal	*Volts*
Gold	+ 1.5
Platinum	+ 1.2
Silver	+ 0.799
Copper	+ 0.337
Hydrogen	0.000
Lead	– 0.126
Tin	– 0.136
Nickel	– 0.250
Cadmium	– 0.403
Iron	– 0.440
Chromium	– 0.71
Zinc	– 0.763
Aluminium	– 1.66
Magnesium	– 2.37
Lithium	– 3.02

Table 7.1 lists the electrode potential which exists between pure metals in the presence of an electrolyte, with hydrogen (which is a metal) taken as zero. A metal higher in the table will be 'protected' by the metal below. The higher metal is known as the cathode and the lower the anode. It is the anode metal which corrodes, passing electrons into solution to protect the cathode in a process known as galvanic action.

This table is obviously of considerable interest to metallurgists and chemists but normal engineering components seldom use pure metals. Table 7.2, therefore, has been found to be useful in indicating corrosion potential between metals and alloys commonly used in engineering, in various combinations.

This table is used by taking the metal considered, which is in the columns, and comparing this with the metal on the left hand side. This gives an indication of the degree of corrosion which could be expected.

The visual effect identified with galvanic corrosion is similar in all respects to the attack found with chemical corrosion. The attack will be more serious close to the anodic part of the metal and less serious remote from this point, that is in the area of action. The further apart the metals in the electrochemical series, then the more serious will be the corrosion of the anodic component.

The galvanic effect can also be used as a means of corrosion protection. Metals such as magnesium, aluminium, and zinc are used as anodes to transfer a coating of metal to the cathode material. Good electrical connection must be made to the components being protected such as bridges, ships, pipelines or other items immersed in water, or damp earth.

A failure here would be where the steel or the metal being protected was corroding because of the wrong placement of the anodes or possibly because an anode was not acting. Visual examination would indicate clearly whether or not anodes were losing metal and providing protection to the cathode. Aluminium in particular has a tendency to form an adherent self-healing oxide which is a non-conductor and thus the galvanic effect cannot proceed after a while. (See figures 7.16 to 7.18).

Table 7.2 Galvanic corrosions: degrees of attack (A=low, C=high)

Metal considered	*Gold, platinum, rhodium, silver*	*Monel, inconel, nickel/molybdenum alloys*	*Cupronickels, silver solder, aluminium bronzes, tin bronzes, gunmetals*	*Copper, brasses, 'nickel silvers'*	*Nickel*	*Lead, tin, and soft solders*	*Steel and cast iron*	*Cadmium*	*Zinc*	*Magnesium and magnesium alloys (chromated)*	*Stainless steel*			*Chromium*	*Titanium*	*Aluminium and aluminium alloys*
											Austenitic 18/8 Cr/Ni	*18/2 Cr/Ni*	*11% Cr*			
Contact metal																
Gold, platinum, rhodium, silver	-	A	A	A	A	A	A	A	A	A	A	A	A	A	A	A
Monel, inconel, nickel/molybdenum alloys	B	-	A	A	A	A	A	A	A	A	A	A	A	A	A	A
Cupronickels, silver solder, aluminium bronzes, tin bronzes, gunmetals	C	B or C	-	A	A	A	A	A	A	A	B or C	B	A	B or C	B or C	A
Copper, brasses, 'nickel silvers'	C	B or C	B or C	-	Bor C	B or C	A	A	A	A	B or C	B or C	A	B or C	B or C	A
Nickel	C	B	A	A	-	A	A	A	A	A	B or C	B or C	A	B or C	B or C	A
Lead, tin and soft solders	C	B or C	B or C	B or C	B	-	A or C	A	A or C	A	B or C	B or C	B or C	B or C	B or C	A

Steel and cast iron (a) (f) (w)	C	C	C	C	C	C	–	A	A	A	C	C	C	C	C	B
Cadmium (u)	C	C	C	C	C	B	C	–	A	A	C	C	C	C	C	B
Zinc (u)	C	C	C	C	C	B	C	B	–	A	C	C	C	C	C	C
Magnesium and magnesium alloys (chromated) (b) (a)	D	D	D	D	D	C	D	B or C	B or C	–	C	C	C	C	C	B or C
Stainless steel: Austentic 18/8	A	A	A	A	A	A	A	A	A	A	–	A	A	A	A	A
Stainless steel: 18/2 Cr/Ni	C	A or C	A or C	A or C	A	A	A	A	A	A	A	–	A	A	–	A
Stainless steel: 11% Cr	C	C	C	C	B or C	A	A	A	A	A	C	C	–	C	C	A
Chromium	A	A	A	A	A	A	A	A	A	A	A	A	A	–	A	A
Titanium	A	A	A	A	A	A	A	A	A	A	A	A	A	A	–	A
Aluminium and aluminium alloys	D	C	D	D	C	B or C	B or C	A	A	A	B or C	B or C	B or C	B or C	C	–

A = The corrosion of the 'metal considered' is not increased by the contact metal.

B = The corrosion of the 'metal considered' may be slightly increased by the contact metal.

C = The corrosion of the 'metal considered' may be markedly increased by the contact metal.
(Acceleration is likely to occur only when the metal becomes wet by moisture containing an electrolyte, e.g. salt, acid, combustion products. In ships acceleration can be expected to occur under in-board conditions, since salinity and condensation are frequently present. Under less severe conditions the acceleration may be slight or negligible.

D = When moisture is present, this combination is inadvisable, even in mild conditions, without adequate protective measures.

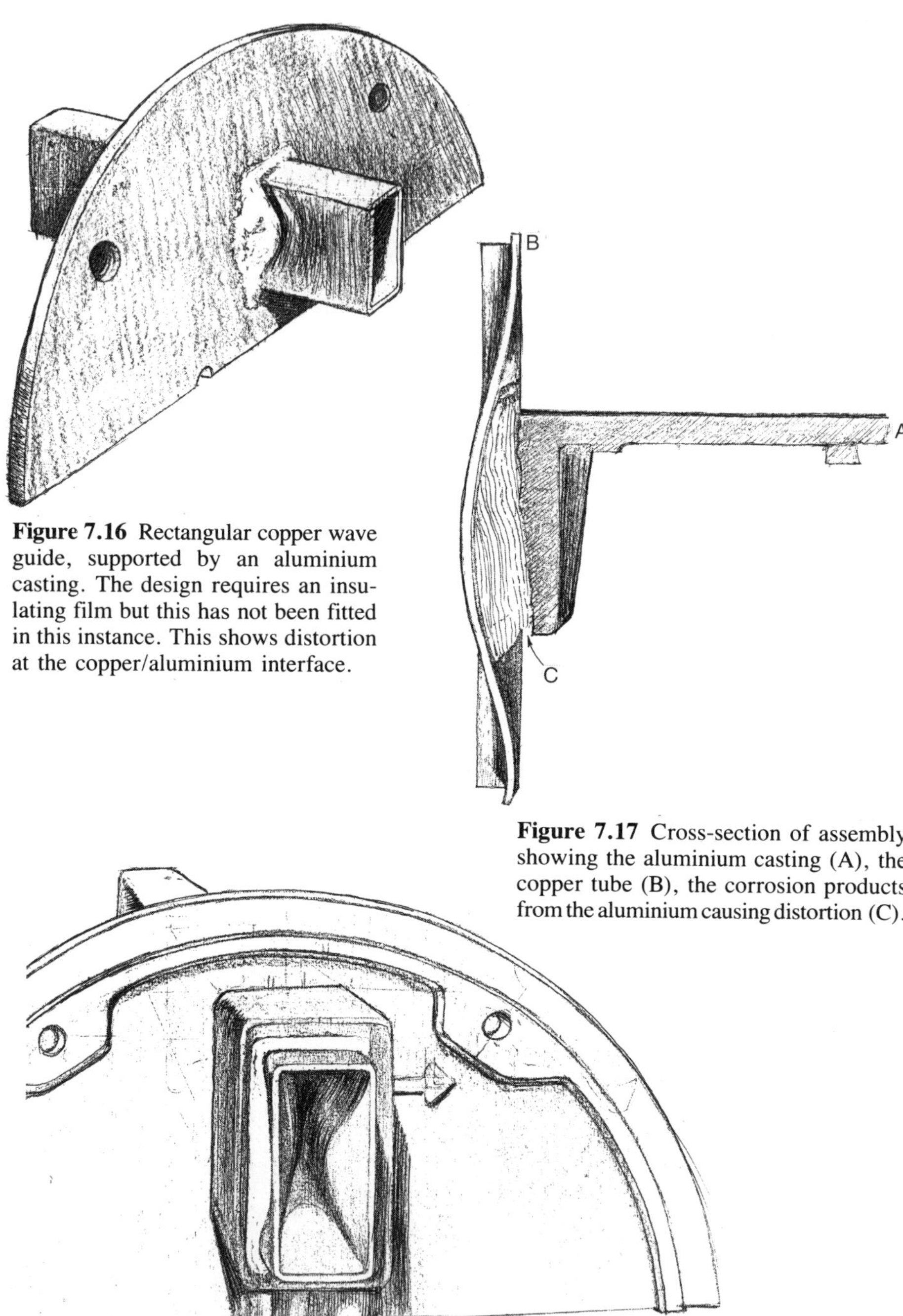

Figure 7.16 Rectangular copper wave guide, supported by an aluminium casting. The design requires an insulating film but this has not been fitted in this instance. This shows distortion at the copper/aluminium interface.

Figure 7.17 Cross-section of assembly showing the aluminium casting (A), the copper tube (B), the corrosion products from the aluminium causing distortion (C).

Figure 7.18 Distortion on the inner surface of the copper rectangular tube.

Electrochemical or electrolytic corrosion

This is basically similar to galvanic corrosion except that the two metals in contact need not be dissimilar but one metal is forced to become cathodic by the passage of stray electric current through the joint. Electrochemical corrosion in this case is the direct result of a DC current passing between two metals. There must also be an electrolyte present which is a conductive liquid. This will result in an electroplating-type cell. The positive metal in this case will corrode, and the negative metal or cathode will be protected.

Visual examination will show typical symptoms if this is the mechanism of corrosion: pitting will be expected to be present indicating metal removal, and the pitting can be bright; there will be evidence of preferential corrosion on points or edges.

Because of the mechanism, it is common that the corrosion is sharply defined but small in area. To an experienced investigator, electrolytic corrosion can be quite positively identified by visual examination alone.

Figures 7.19 to 7.21 illustrate the effect of electrochemical corrosion.

Micro-examination, if carried out, will show little or no intergranular attack. That is, the metal is removed more or less homogeneously whereas with galvanic and chemical corrosion the mechanism is for attack to occur around the grain boundaries with the grains falling out.

In order to cure the problem it is desirable that the source of the electric current is identified. On occasion this can be quite simply carried out by using a direct current voltmeter between the metals which are involved in the assembly. If this shows voltage differences greater than 10 millivolts then this will be the source of the problem, with the corroding metal being positive. Experience shows that the voltage causing the problem is often intermittent, and can be difficult to pinpoint. It is then necessary to install a recording voltmeter in the system which will have a chart recorder to show when voltage

Figure 7.19 Local well-defined corrosion pits on copper. This was in contact with steel which was not corroding, clearly indicating that electrolysis must have existed under the influence of a stray electric current.

is applied and from which it may be deduced that some electrical equipment had been switched on at the relevant time. Further investigation can then be carried out by a competent electrician. Static electricity can be a potential cause of this corrosion.

Electrolytic corrosion can almost always be controlled or eliminated by adequate earthing, but it is obviously preferable to identify the source of the problem.

There are case histories which report the problem to be the galvanic corrosion/protection running out of control, or poor earthing of electrical equipment. It is not always appreciated that alternating current can be easily rectified to have a direct current content. This may be the result of poor earthing. There is a history of welding causing problems, when the welder's

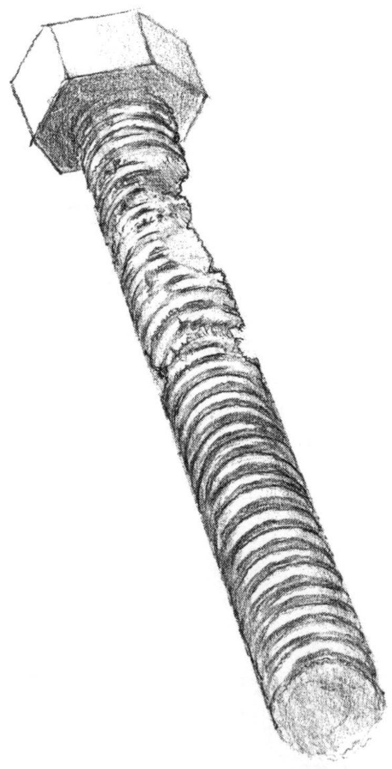

Figure 7.20 A low alloy, high tensile threaded stud with local attack on one side of the threads close to the head.

earth connection has been badly fitted, or connected to equipment not correctly earthed.

Micro-specimens should be prepared from areas which show corrosion. The investigator will also want to see specimens from areas which might have effects such as pitting, scaling or other visual effects to a lesser degree than the problem area itself. It would be usual for at least one specimen to be taken from a surface or surfaces showing no attack. The specimens would be examined first in the unetched condition. This would show the degree of surface oxidation, whether this was adherent, coherent, in layers or intergranular. The etched structure would show the extent of any intergranular attack, and also the metallurgical structure.

The investigator would use this information to evaluate whether the corrosion was chemical, galvanic or electrolytic. This would be based on the type and extent of the attack, particularly the grain boundary effects, and requires considerable experience. It is not uncommon to find that serious corrosion is the result of all three types.

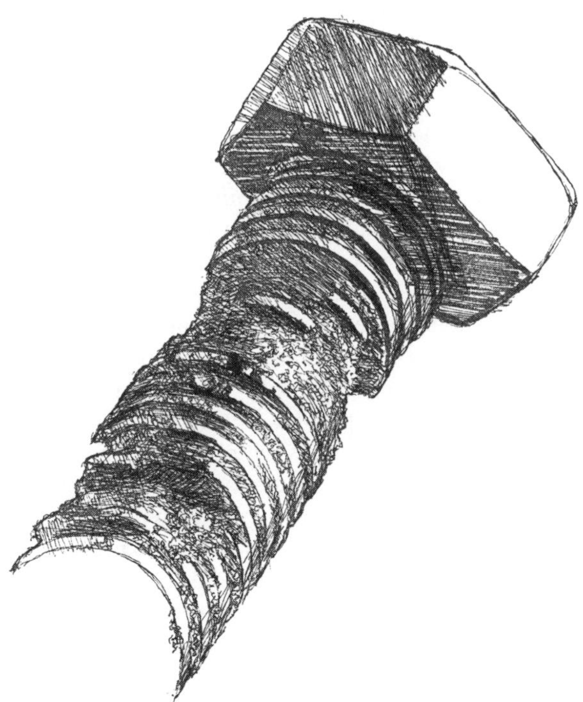

Figure 7.21 Close-up of the area under the head of the component in Figure 7.20 showing the sharp demarcation of the corrosion. This component was found to be at 120 mV (anodic) with respect to earth, indicating electrolysis had taken place.

General comments on corrosion/contamination/degradation

This is a visual effect which can result in an investigation because the component is no longer 'fit for use'. In addition, however, corrosion and contamination can be the initiation of failure which propagates by different mechanisms. This could be through fatigue, brittle tensile stress or abrasion.

Corrosion, for example crevice corrosion, pitting, stress corrosion or where grain boundaries are attacked, results in acute fatigue stress raisers. The fatigue will be identified in the visual examination as the mechanism of failure, with corrosion on further investigation as the reason for failure. The term 'corrosion fatigue' is commonly used to describe this, but is also loosely used, and care is required in having the term defined. This book does not recognize or use the term 'corrosion fatigue'.

Crevice corrosion

Crevice corrosion is a term used for pitting corrosion as the result of the surface being covered to prevent oxidation. Crevice corrosion problems are most commonly identified with stainless steels but can occur with all metals where a homogeneous oxide can be produced to protect the underlying metal. Where this oxide is destroyed, as with self-healing metals such as stainless steel, aluminium, titanium, nickel, and to a lesser extent copper, the oxide will re-form in an oxidizing atmosphere. Where, however, the surface is covered for any reason, such as with a lap joint or with contaminated material, and is then abraded to remove the oxide, the surface will not be re-oxidized. A corrosive liquid or atmosphere can then attack the underlying metal, and this corrosion will propagate in the form of a pit.

Pitting corrosion is sometimes found in local areas where the corrosive environment for some reason is increased. A good example of this is the severe local – generally pitting corrosion – found at the base of steel/cast iron railings, which remain damp. This pitting on mild and low alloy steel is less easy to explain than with metals which form cohesive oxide coatings. It would appear, however, that the same theory applies, that is if a homogeneous oxide film can be formed and retained, pitting will not advance.

Scaling corrosion of mild steel, that is the formation of a surface layer of iron oxides, occurs under specific conditions, and very often seems much more serious than it really is. It would appear that the original surface corrosion must be porous, with the corrosive conditions being specific. For example, steel sheets stored on top of each other in moist/damp conditions. This allows the corrosion to continue on the surface. The oxides formed have a volume 5–7 times greater than iron which appear to seriously affect the component size when removed, whereas the actual loss of metal is quite small.

If this oxide layer is produced under controlled heating conditions, for example, hot rolling or heat treatment, it can be adherent and supply a high standard of corrosion resistance if it is left *in situ*. The very subtle differences between corrosion and protection are shown by the severe pitting at the base of mild steel railing posts, with no corrosion 100 mm above.

There are case histories where mild steel components have improved pitting corrosion resistance from the oxide film produced at heat treatment.

There is a strong argument for an investigation into why some mild and low alloy steels resist corrosion under apparently exactly the same conditions which cause severe corrosion in others.

Crevice corrosion pitting can be severe and very rapid where aggressive liquids are used, as with, for example, stainless steel or aluminium which may have been chosen because of a corrosive atmosphere. Crevice corrosion, once identified, can be cured by cleaning and re-oxidizing the surface using a hot flame or oxidizing liquid. It can be prevented from recurring by ensuring that the surfaces are kept clean and not covered. Crevice corrosion can result in failure by leakage and thus is identified as a mechanism of failure and reason for failure. It can also be the initiation of fatigue failure. Pitting (crevice) corrosion can be in the form of rust or bright pitting.

Stress corrosion

Stress corrosion results in cracking from a combination of corrosion and tensile loading. At the time of writing, failure from stress corrosion is the most common mechanism and cause of failure. This can be explained by design engineers operating materials further up the stress/strain curve.

This failure mechanism has a very typical pattern of bifurcation or multiple cracks and is readily identified at visual examination when severe as in Figure 7.22. Stress corrosion is caused when a material is in an environment which produces a surface layer because of the reaction of the metal with the atmosphere. This layer will always be more brittle than the underlying metal and thus at low levels of movement (stress) can be expected to crack. The underlying metal will then proceed to corrode, again producing a brittle surface.

Because the surface layer is extremely brittle the normal propagation of a crack from the nose does not occur and side wall cracking takes place instead. This results in the bifurcating cracking which is typical, and unique, to stress corrosion.

It must be appreciated that oxide-filled bifurcated cracks can and do exist as surface effects on 'as forged' and 'as cast' surfaces. These normally have characteristics which will be identified by the competent metallurgist as being different from stress corrosion, and would not propagate in the same form, but might act as stress raisers for fatigue propagation.

It is not uncommon for fatigue to be identified as the mechanism of failure, and for micro-examination to identify stress corrosion cracking as the reason for failure. It would, however, appear that there are investigations which miss the minute stress corrosion cracking and report the failure as the result of fatigue.

It would also appear that stress corrosion cracking (SCC) is no longer considered a practical problem to solve, but is embroiled in the theoretical aspects of hydrogen ion diffusion and embrittlement.

The author is convinced that when stress corrosion is identified, which is not difficult for the experienced practical metallurgist, it should be possible to advise on a cure.

Stress corrosion then is the result of stress plus corrosion. It is one of the rare occasions when the metallurgist has managed to use simple English to

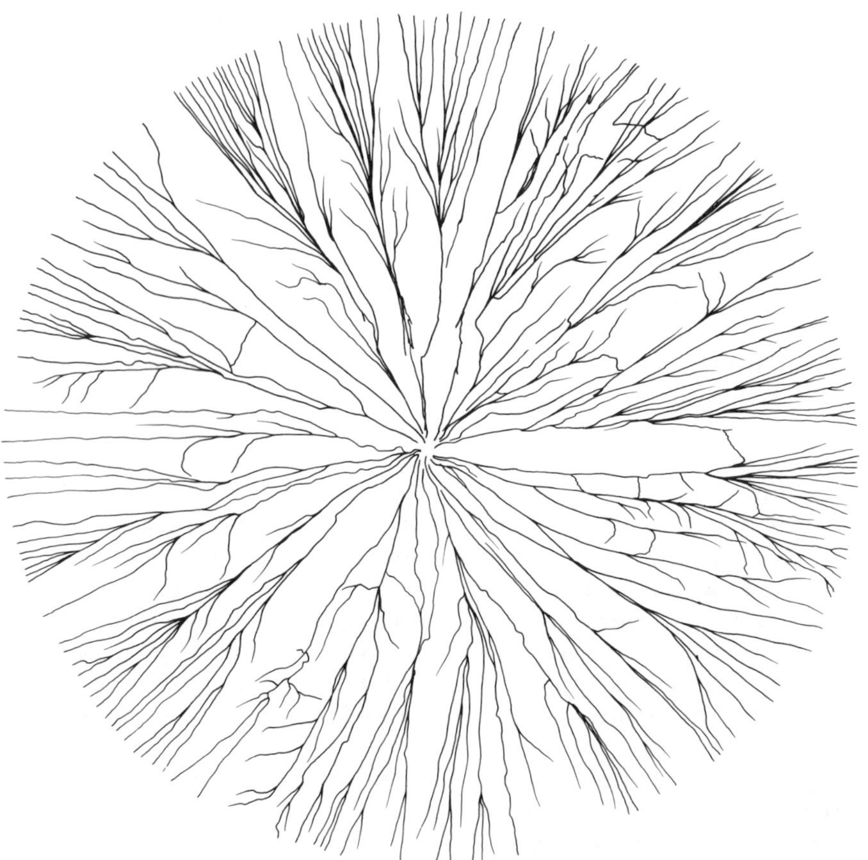

Figure 7.22 Stress-corrosion cracking. This is an austenitic stainless steel sheet with a light tensile load applied to the centre and with chloride contamination. It shows the typical bifurcating cracking.

describe a metallurgical feature. Where it is confirmed the investigator has four options to cure the problem.

1. Reduce the stress – with the same corrosion environment. This would appear to be the route of NACE (with their Rockwell C22 maximum hardness).
2. Eliminate or reduce the corrosive environment.
3. Increase the yield/tensile strength of the material involved.
4. Improve the corrosion resistance of the material used.

To enlarge on each of the above:

1. Reduce the unit stress That is, ensure that the stress level in service is below the critical level where corrosion is relevant. The NACE (North American Chemical Engineers) specification requires that no component which might be subjected to stress corrosion conditions shall have a hardness in excess of Rockwell C22. This restricts the design engineer to steels with an ultimate tensile strength of 650 N/mm^2 (45 ton/in^2). The estimated yield strength of such steels would be in the region of 400–500 N/mm^2 (25–30 ton/in^2). This means that the designer must alter the engineering conditions of operation, or use larger sections.

The puzzle remains that many components designed for use at a hardness of Rockwell C30 or above (ultimate tensile strength 1000 N/mm^2, estimated yield (proof) strength 750 N/mm^2) have been taken out of service, tempered to reduce the hardness to an estimated yield strength of 500 N/mm^2

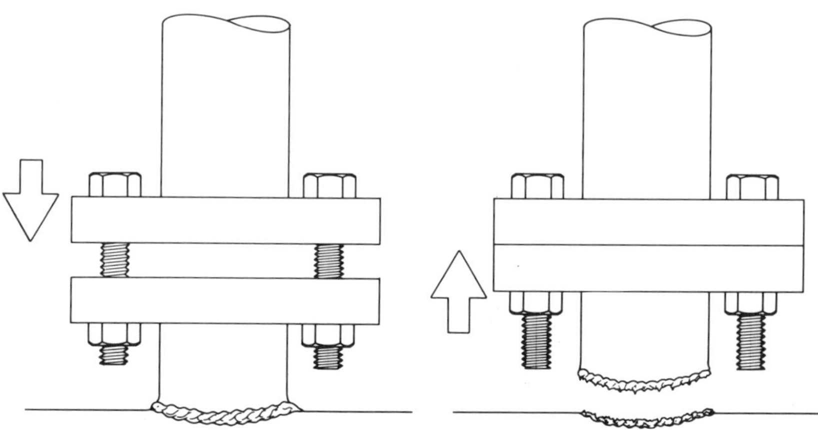

Figure 7.23 Weld joint fracture due to tension caused by a gap between flanges. When a gap occurs which is closed by tightening the bolts, a tensile load is applied to the tank surface.

(Rockwell C22), and then returned to service with no history of failure from fatigue or tensile stress.

There can, however, be situations where with an existing design the operating conditions can be amended to reduce the stress. A dramatic example of this is the use of stainless steel tanks with steam heated jackets in the brewing and distilling industries. These tanks had a history of steam leaks which were shown to be the result of stress corrosion cracking from chloride contamination (corrosion) with a distinct pattern, identified as a tensile pull (stress) at the steam inlet.

This has been cured by ensuring that when fitting the steam inlet and outlet flanges, these are pushing – compressing – the jacket. That is, the stress element was reduced without affecting the corrosion. Thus, any tensile load on the jacket is reduced or eliminated, the corrosive component remains as before, the stress is removed and the problem cured (see Fig. 7.23).

2. Reduce the corrosive environment This is seldom a practical consideration. Ships operate in a marine environment, laundries have a hot humid atmosphere, neither of which can be readily modified.

There are examples of altering the environment, but these are the exception rather than the rule.

3. Increase the material strength There may be occasions when an increase in yield strength will stiffen the component, thus reducing the movement which cracks the surface layer. This will be of secondary significance and is not a practical suggestion. The failure investigator can tell from the metallurgical evidence the ratio of stress to corrosion, and may be able to advise if an increase in stiffness could be a possible cure.

4. Improve the corrosion resistance This is probably the simplest, and thus most economical, technique. There are examples of crevice and stress corrosion problems which have been cured by improving the corrosion resistance.

It is obviously difficult, if not impossible, to be positive that a cure has been achieved, but there is evidence that by cleaning stainless steels (austenitic and 12% chrome) and then oxidizing the surfaces by heating or chemical treatments, crevice corrosion and stress corrosion have been eliminated.

The use of surface treatments, oil, phosphate, zinc electroplate, hot dip galvanizing, sheradizing, nickel plating or painting can be successful providing they improve surface corrosion resistance. The 'safe' way to prevent stress corrosion is to specify sophisticated corrosion-resistant steels.

The straightforward 18/8 chrome–nickel and 12% chrome–stainless steels have been shown to be susceptible to stress corrosion cracking. There are now many modifications of these steels, notably the Duplex varieties, which have significant resistance to stress corrosion and crevice corrosion.

Use is now being made of copper alloys such as beryllium copper and cupro-nickel, nickel alloys such as Monel, and in extreme cases 80/20 nickel–chrome alloys. Titanium and its alloys are potential candidates waiting in the wings for a reduction in the cost of titanium. Because of the temperature limitation of approximately 200–300°C maximum there will always be rather limited use of titanium.

There is little doubt that the debate and research into SCC (stress corrosion cracking) will continue. It is to be hoped that this can be removed from the esoteric into the practical aspects of the problem.

Biological corrosion

One source of corrosion which can be serious is biological, either from natural growth or from the presence of bacteria in the atmosphere or liquid in contact with a metal. This can produce a change in pH, generally to make it acid but on some occasions can increase the pH thus attacking aluminium or magnesium. More common and more serious is the effect of bacteria. It is now known that there are certain types of bacteria which can produce sulphides and sulphates in relatively large quantities and high concentrations. These can cause serious problems regarding corrosion of almost all metals and are particularly serious with nickel alloys.

It is therefore essential that the investigator appreciates these possibilities, discusses, asks questions and samples the atmosphere to identify the source of such problems.

There have been instances where bacteria exist under certain conditions causing corrosion, but when the conditions alter the bacteria disappears, leaving corrosion from no obvious source. This can occur with moisture contamination. There is a history of corrosion problems in fuel tanks where moisture was present from condensation in the tank or supply system, on an infrequent basis.

Bacteria and fungal growth can be encouraged under some conditions of temperature and contamination, resulting in the formation of aggressive corrosive conditions. There is evidence that this can initiate corrosion, but then be flushed out leaving the corrosion to propagate for no apparent reason.

Another problem is iron. Bacteria can concentrate the small amount of iron present in water, producing iron oxides which can block filters, and in contact with other metals give galvanic cells. This will only occur where water is stored in tanks or is more or less stagnant.

Contamination

Contamination will be identified by visual examination, commonly following a non-conformance. For example, the blocked filter above would be the result of contamination, which on examination could be the bacteriological effect. This can result in fuel blockage in diesel engines.

With concrete and bricks contamination can be a problem resulting in efflorescence which may be a visual problem only. Further investigation would then be necessary to evaluate whether or not there could be chemical attack.

Sulphur compounds such as sulphite, sulphate and sulphide can cause problems with concrete and brick. Carbonates and chlorides can cause efflorescence, and may or may not cause problems, but should be removed by washing and brushing.

Wood products are susceptible to damp and heat, bacterial attack and fungal growth resulting in visual contamination. When found at an early stage remedial action can be taken, otherwise rot from the surface will propagate. Wood can also fail from the centre if the tree from which the wood was produced had irregularities. This would be classed as degradation.

With plastics/polymers, contamination in conjunction with high humidity and temperature can result in rapid degradation. Also, many solvents used for cleaning metals will attack the surface of plastics resulting in softening, material removal and in some cases degradation of the base material. Contamination of plastic/polymer surfaces can alter chemical make-up resulting in a brittle layer which could cause brittle fracture.

Plastics/polymers, wood, and to a lesser extent cement, mortar or concrete, are subject to degradation where time, light, normal environment and temperature result in chemical changes. These may be seen under some circumstances, but this is not usual. The result of degradation can show up as dimension change, but more generally it is found by change in mechanical strength, most commonly brittleness. This in conjunction with dimension change may be seen as surface cracking.

It is, however, probable that degradation due to contamination is most commonly identified by brittle fracture. Thus the mechanism of failure is brittle tensile stress or surface cracking, and the reason is degradation. This can be identified by modern chemical analysis provided there is available either a detailed specification or more usefully an example of a component not subjected to the environment of the failed or problem component.

Prevention of corrosion

During discussion on the above in various other chapters such as Chemical analysis and Materials information, techniques for corrosion protection or prevention are dealt with in more detail. Here we discuss the techniques available.

Corrosion/contamination/degradation are the result of contact with the environment. If the environment is a high vacuum at room temperature or below, then few problems will exist either with metals or non-metals. This is obviously not a practical solution except in extreme cases.

Where the environmental temperature is low, that is below say −20°C, then corrosion problems on metals will not exist and it is unlikely that degradation of plastics/polymers will take place except extremely slowly. Contamination, however, can remain a problem, but again it is not a practical proposition outside the Arctic/Antarctic to reduce the temperature to prevent corrosion/contamination/degradation.

Most components used by industry operate in an oxidizing atmosphere which is breathable or, under some circumstances, in atmospheres or environments which are corrosive because of the contents or conditions in which they operate. Thus for the majority, the environment is a problem.

With corrosion protection, price will almost always be a significant factor in the design equation, even with nuclear, petrochemical and suchlike industries. There is seldom a suggestion that platinum or gold plating could be used although this could be the ultimate in corrosion protection. Price will always have a part to play when the designer starts with a clean sheet of paper. It would be wrong for a designer to specify high tensile titanium for a component with an expected system life of 10/15 years at ten times the cost of some other cheaper material correctly corrosion protected. This would apply in particular if the component can be replaced economically after say every 5 years.

The remainder of this chapter will take the sixteen groups of materials listed in Chapter 13 on Materials information and discuss how they could be corrosion protected. (Part 4 describes why they would be chosen in the first place.)

Iron and its alloys

This can be divided into cast iron, mild steel and alloy steel. These corrode in the normal environment (that is the one which is breathed) especially when temperature rises and pH drops, in the presence of moisture or high humidity. Contamination from chlorides, nitrates, etc. will accelerate corrosion. Then there is stainless steels which require an oxidizing atmosphere to prevent corrosion.

In passing it should be commented that wrought iron (which is no longer considered for use by engineers) has an excellent corrosion resistance history, but low strength. It is no longer available in economical quantities. Cast iron also has excellent corrosion resistance.

To return to the normal iron alloys; these do not corrode in the cold, or in a hot dry atmosphere. A thin film of oil will supply further protection under many circumstances. They can be protected to some extent by controlled oxidation. The black oxide process is the modern equivalent of 'blueing' of gun barrels. This uses hot caustic solution to which is added controlled oxidizing agents such as chromates. This was a development from the 'blueing' carried out by gunsmiths which was a very complicated secret process with things like ants eggs and spiders webs! The modern gun is treated with a form of black oxide which has limited application but is still used with oil for some tools where there is adequate control and the personnel are skilled and trained.

Chromate treatment of clean steel will give limited corrosion protection. Phosphating is a common method of protection where low level protection is required. With this, the surface of the metal is converted to iron phosphate. This in its own right has better corrosion resistance than iron and the phosphate tends to be more adherent than iron oxide. It also has the considerable advantage that it acts not unlike blotting paper in that it can absorb liquids into the thin layer of phosphate. This can be oil which will give some form of temporary protection but obviously does not aid handling. Resins, however, can be absorbed and are more permanent.

Phosphating is used as a preparation for painting. Almost all modern motor car bodies are phosphated, followed immediately by a paint which is absorbed into the phosphate and then stove dried. A phosphate coating sealed by chromate treatment increases the corrosion resistance, but can reduce the paint adhesion.

The purpose of black oxide, phosphate and chromate is to convert the surface into something more adherent than iron oxide.

Painting is probably the most common method of treating mild and low alloy steels. The traditional paint was based on linseed oil with metallic pigments, mostly oxides. Linseed oil has the characteristics of oxidising to change from a liquid into a solid and thus the paint could be applied as a liquid, left in the atmosphere and became solid. The pigments in the original paint were almost invariably for decorative purposes although in some cases they supplied corrosion protection.

Linseed oil traditionally was thinned with turpentine which evaporated. The original paint therefore had three components:

- A liquid which converted into a solid or semi-plastic solid at room temperature. This was known as the 'vehicle'.
- A pigment which supplied colour, and where necessary body.
- A thinner which allowed the paint to be applied more readily and then disappeared into the atmosphere.

The colour range in the early days of painting was quite limited and chemical problems were found to exist. For example, the green pigment was almost invariably an arsenical compound, and there is some evidence that this killed Napoleon because his bedroom was painted green!

The red pigment commonly used was the red oxide of lead, and this is now known to be highly toxic. This was found to be an excellent primer for corrosion resistance, and was used as a primer until quite recently. There is, however, a history of serious health problems associated with the application of red lead, and also for those who were involved in salvaging components. The red oxide of lead is no longer used.

Modern paints have advanced from the linseed oil base to plastics/polymers which convert from liquid to solid, sometimes at ambient, sometimes after heating, and at other times requiring an additive stirred in prior to use, known as an accelerator.

There has been a continuing change from metallic oxide pigments to organic particles. Modern pigments are almost invariably plastics/polymers with excellent resistance to the effect of light, using stable organic pigments. It is on record however, that until the quite recent past companies such as Rolls Royce would not supply a red car because of the poor stability of the red pigment.

Pigments are used for three distinct purposes, not all of which will always apply:

1. To supply bulk to the paint. This achieves the thickness required either for decorative or corrosion protection purposes.
2. Probably most important technically, pigments can supply positive protection by their galvanic action. These pigments contain aluminium or zinc. Aluminium is lower on the galvanic scale than zinc, and thus in theory at least will give a more positive protection. Aluminium, however, tends to form a protective oxide which inhibits the galvanic effect as this is a non-conductor. Aluminium should not be used as the pigment where the material has already been protected by zinc, either through electroplating or galvanizing. In theory at least, the aluminium will sacrifice itself to protect the zinc, and this is not as useful as the zinc sacrificing itself to protect the steel. There are now ion exchange pigments which are being found to give positive corrosion protection, using free ions to give the galvanic effect. These are quite modern and not in common use.
3. The pigment is used to give a colour either for decorative purposes or for identification. Where more than two coats are specified as part of the protection system, then it is a very simple technique to specify three different colours. Thus the painter, paint inspector and the client can readily identify that the three coats have been applied.

It must be appreciated that it is surface preparation which is the single most important aspect of paint application. This ideally should be a blasted surface with the surface contours being carefully controlled to give a specified height to depth ratio and with smooth contours free of pinnacles. This ensures firstly that the paint has a key for adhesion, and secondly that it does not drain off sharp peaks which would give a very thin coat on the apex and much thicker at the base.

The primer, which for best protection should have some positive corrosion protection attributes, must be applied as soon as possible after blasting. Four hours would be the ideal delay, with no longer than 12 hours. The surface must be kept dry and preferably warm between blasting and painting.

Where necessary a primer and 'tie' coat should be applied when for any reason, such as transport damage, the final coat cannot be part of a continuous paint system.

The life expectancy of the satisfactorily coated component will depend on a variety of factors, these being firstly the environment. If this is aggressive such as a warm, marine atmosphere with changes in temperature then the protection will be much less for the same type of coating than for a component to be used in a cold stable climate with low moisture.

Whether or not the surface is damaged will of course be of more significance than the type and thickness of coating.

There is now information available from reputable sources regarding the type of application, that is primer, tie coat and top coat, the total thickness of each coat to be expected for different types of coating for multiple applications. It is unlikely that the failure investigator will have this information in sufficient detail to satisfy all potential clients. It would, however, seem preferable that the investigator who has identified the problem and is aware of the environment should be involved with establishing the specification. The investigator usually has a better technical ability to discuss with applicators, paint suppliers, and corrosion consultants, than the client. It can be quite difficult for the client to differentiate between a genuine consultant who has no axe to grind regarding the type of paint thickness, etc., and the 'consultant' who has some tie-up with either a paint company or applicator.

It is now possible to specify and ensure that the specification is correctly applied to achieve a long life, certainly 20 years, and up to 50 years, of successful resistance to corrosion from the normal environment. The Forth Railway Bridge is 100 years old and has been painted approximately 99 times. On the other hand the Forth Road Bridge is approximately 30 years old and at the time of writing is being touched up locally with, as far as the author is aware, no plans for continuous painting.

Forms of corrosion protection for steel would be zinc, cadmium, and to a lesser extent aluminium, nickel, tin, and lead. Zinc deposits are by far the most common, there being four types – electroplate, hot dip galvanize, sheradize and metal spray.

Zinc electroplating has a deposit thickness of approximately 0.01 mm (0.0005″), galvanize and sheradize have deposit thicknesses of 0.1 mm (0.005″), and metal spraying can have any desired thickness. These are all galvanic coatings, and have better resistance to the environment than the mild and alloy steels.

Corrosion of zinc-coated components is in three stages.

Stage 1 This is the environmental chemical attack on the surface. There are now statistics supplied by the Zinc Development Association giving the rate of corrosion of zinc under different conditions. These vary from the very low corrosion rate in the Arctic, to medium rate under inland temperate conditions, with increase of corrosion in industrial, marine and tropical conditions, as in Table 7.3.

Table 7.3 Corrosion rates of zinc

	Loss of zinc	
	mm per year	*inches per year*
Severe industrial	0.013	0.0005
Coastal marine	0.0057	0.0002
Industrial	0.0054	0.0002
Rural	0.001	0.00004
Rural marine	0.005	0.0002

The atmospheric temperature will have a significant effect, reducing the corrosion with low temperature, increasing corrosion as the temperature rises.

The resistance to chemical attack can be further increased, particularly with electroplated zinc, by chromate sealing. It can be shown that the rate of attack on the zinc is proportional to the atmospheric conditions, but that with a correctly applied chromate seal the onset of this attack can be considerably delayed or prevented altogether.

The first evidence of zinc corroding is the presence of a white deposit sometimes referred to as 'white rust'. Once this appears it is a matter of time until the coating is destroyed leaving bare metal.

Stage 2 This is when the galvanic protection will commence, with the zinc sacrificing itself to protect the underlying iron alloy. This will destroy the zinc coating more rapidly than the chemical attack.

Stage 3 Once there is, however, a large enough area, this iron alloy will rust (red rust) with the zinc continuing the battle to protect the metal in contact with the zinc. The failure investigator will have identified the onset of the red rust, and with a modicum of luck will be able to advise if the problem is damage, chemical attack, poor quality chromate, or too thin a deposit of zinc.

The above comments apply to all forms of zinc protection. It can, however, be expected that each form of protection will have been applied for specific purposes or components.

Electrodeposit This is used on a large scale for electrical, electronic and household articles, etc. Most electroplated items are 'passivated' using chromates.

Many components are now painted after zinc plate and chromate. This requires a special active primer to ensure adhesion of the paint.

Electroplate (electrogalvanize) is commonly used on studs in preference to hot dip galvanize, but will never have the same corrosion resistance. There is a common history of corrosion on these components where they are used in conjunction with hot dip galvanizing. The electrodeposit is never thicker than 0.02 mm (0.0008″) and can be as thin as 0.002 mm (0.00008″).

Hot dip galvanize This is a traditional process where the components are cleaned, pickled, and dipped in molten zinc. This can have additives of aluminium to improve the appearance, and is sometimes chromated to seal the surface after cooling.

The coating thickness can vary dramatically depending on the configuration, method of dipping, cooling, etc. and will seldom be less than 0.075 mm (0.003″), but can be as much as 0.3 mm (0.01″). This applies to mild and low alloy steels where this thickness precludes the use on any component with accurate tolerances, such as bolts, studs, nuts, etc. They may be machined after galvanizing but this can cause problems, and sheradizing should be specified for such pieces.

Sheradize This is a diffusion process where the components are packed in sealed bags with activated zinc granules heated to approximately 400°C. This produces an interface of zinc/iron alloy, with a coating of zinc with some zinc oxide. The coating thickness can be compared to hot dip galvanize, but follows the surface contours, and thus is suitable for nuts, bolts, etc.

Metal spray Zinc is now commonly applied by conventional metal spray. The coating thickness and homogeneity will largely depend on the skill of the operator. Metal spraying is generally confined to large components and structures, which are very often subsequently painted.

Other processes for steel finishing There is also aluminizing where the steel surface is given an aluminium/aluminium oxide coating. This can be applied by immersing the blasted steel in molten aluminium or by metal spraying the component followed by heating, or chromating to oxidize the aluminium. Components for 'furnace furniture' for use up to 500–600°C are commonly aluminized.

For long term applications – up to 50 years, corrosion-free life steel – structural steel can be blasted, aluminium sprayed or immersed in molten aluminium, then coated with high quality paint. Routine inspection will always be required for damage, etc.

Nickel has excellent corrosion resistance and was at one time commonly used in industrial and marine conditions. It will seldom be met with on modern components. It was used for aggresive chemicals, hot oxidation conditions, etc. but not where sulphur or sulphides were present. Stainless steels have largely replaced nickel plating.

Tin plating was commonly used (tin plate, tin cans) for food containers. Tin is non-toxic and provided the plating deposit was not porous, no problems existed. While still used to some extent, it is being replaced by aluminium 'tins', or plastic bottles.

Lead plating for corrosion resistance will be seldom encountered by the failure investigator. It may be found where aggressive chemicals are used, and provided it is not damaged or porous, a thin deposit will resist attack from most acids.

The above comments apply to mild and low alloy steels, where the main problem if left untreated is that they tend to form an oxide which has low adhesion which readily falls off. This does not apply to either austenitic stainless steels plus additives or 12% chromium steels. While these are both termed 'stainless steel', they have different characteristics. The austenitic stainless steels, that is 18% chrome, 8% nickel, are single-phase materials. That is, they do not have any galvanic cell action between the internal constituents. This makes them similar to pure metals which invariably have good corrosion resistance. In addition, the surface oxide is chromium, not iron. This is adherent and self-healing. This dramatically improves their corrosion resistance against firstly the atmosphere, but particularly against aggressive chemicals. Unfortunately, it leaves them with poor mechanical properties with a yield/proof strength less than that of mild steel.

The 12% chromium steels have two or more phases, similar to alloy steel. They do, however, like austenitic stainless steels, have an adherent chromium oxide on the surface which is self-healing. Thus when the surfaces of austenitic or 12% stainless steels are abraded, broken or damaged in an oxidising atmosphere, they immediately re-form the chromium oxide which is adherent. These materials therefore do not rust in the same way as mild and low alloy steels, and are not prone to pitting or scaling provided the atmosphere is oxidizing.

This highlights a problem which is not commonly appreciated by many people involved in writing specifications, that is that austenitic and 12% chromium steels should not be used in an atmosphere which is reducing, i.e. not oxidizing.

These steels should not be obscured by designs such as lap joints or be covered by an 'O' ring for example, on a surface where corrosive conditions exist. There is on record an 'O' ring which was used under vibrating conditions in sea-water which corroded, abraded or trepanned a circle the same diameter as the 'O' ring through 10 mm of the austenitic stainless steel.

If a stainless steel or even a plastic clamp is used in contact with a vibrating pipe then crevice corrosion can be expected. Crevice corrosion, given the correct conditions, can penetrate a relatively thick wall and thus allow the escape of the contents of the pipe.

Where clamping is required then the stainless steel pipe must have an adherent plastic collar either applied by adhesive, or by the application of films of paint/plastic. These must be thick enough and adhesive enough to ensure that no relative movement exists between the internal surface of the film and the stainless steel, and that they will not be abraded away in service.

Austenitic and 12% chromium steels can have the surface chromium content reduced by heating, either welding, heat treatment, or in some specific cases chemical action.

When the chromium content of the 18/8 stainless steel falls below approximately 12%, and the 12% chromium steels falls below approximately 10%, then the self-healing chromium oxide will not re-form. The presence of this low chromium content can be identified with the 'ferroxil test'. This is a sensitive test for iron. It uses activated potassium ferricyanide which turns brilliant blue in the presence of free iron. This can result in the formation of red rust, particularly on 12% chromium steels, or the attack on the surface of 18/8 chromium–nickel steels where the contents of the austenitic steel vessel can become contaminated. This was identified particularly in the whisky industry where it was found that when austenitic stainless steel tanks were welded, free ferrite formed associated with the welds, and the whisky attacked this, turning green. Green whisky is not an acceptable commodity!

The problem is now overcome by 'passivating' stainless steel vessels after welding. This is achieved using controlled acids, or paste, and is in fact 'de-activating not 'passivating'.

The painting for decorative purposes of stainless steels of any type should be resisted.

The Duplex stainless steels which are based mostly on the 12% chromium with additives require special care. Like the other stainless steels they can have the surface denuded of critical elements resulting in local corrosion. The wrong heat treatment can also result, in theory at least, in the corrosion resistance properties being reduced.

As stated elsewhere in this book, there is a history of over-specifying of these materials and to date little evidence of any real problem.

All of these steels, that is 18/8 austenitic, 12% chrome and Duplex, can have their corrosion resistance properties improved under many circumstances by making sure that the surface is cleaned, preferably by wire brushing with stainless steel or emery paper, removing all the debris and then oxidizing with an oxyacetylene flame or, in some cases, chromate solution.

The indications are that the oxidizing flame correctly carried out will enhance the corrosion resistance and can overcome the problems of crevice and stress corrosion on 18/8 and 12% chromium stainless steels under certain circumstances.

The investigator who is aware of what caused the problem will have to decide on the best method of overcoming it. An oxidizing flame can be one technique.

Aluminium and its alloys

Pure aluminium has excellent corrosion resistance against the normal environment. It has no history of corrosion except under alkaline conditions.

It must however be appreciated that aluminium is quite low in the electrochemical series, and will become the anode when in contact with zinc, steel (iron), copper, etc. in the presence of moisture (the electrolyte). The fact that aluminium readily forms a strong oxide which is a non-conductor is an advantage as this inhibits the galvanic effect.

Aluminium alloys, depending on the alloy, will have less corrosion resistance than the pure metal and may require protection. The investigator will need to identify the type of corrosion involved and will then have to decide on the type of protection. The minimum protection normally specified comes under the generic term of 'Alochrome'. This is a proprietary name used in a similar manner to 'Hoover', covering the chromate salts which are used in the process. Properly applied, this will form a stable oxide on the surface and this oxide will have chromates absorbed into it to give it a characteristic yellow colour. Provided this is not damaged, it can withstand all normal environmental problems and while it might have a lesser life under marine conditions, there is evidence that it can give reasonable protection.

Where, for any reason, aluminium requires to be painted, then it must be 'activated' and chromating/Alochrome is the most popular and economical method. While roughening the surface by etching or blasting might be possible, this normally would be more expensive than chromating and not as efficient. Without some form of pre-treatment adhesion of any except very specialist paints to aluminium can be expected to be poor.

Where corrosion resistance against a more aggressive atmosphere is required then anodizing will be specified. This process uses sulphuric acid, or occasionally chromic acid, where the components are electrically connected to the anode and are then subjected to a direct electric current. When first immersed it will be seen that a current passes between the anode and the cathode. As the process continues, however, there is a build up of aluminium oxide on the surface, and being a non-conductor, eventually no current will pass. The higher the voltage used then in general the better the anodic film produced.

There is also a process known as hard anodizing or refrigerated anodizing whereby cooling the solution of sulphuric acid, a much harder, more adherent film is produced. This will not normally be used for corrosion resistance except under very adverse conditions. More information is given in Chapter 13 on Materials information.

While anodized films without doubt can be painted and in fact will act as a key for painting, it would be most unusual for aluminium to be both anodized and painted.

Copper and its alloys

These on the whole have inherently good corrosion protection. Until the advent of stainless steels, they were the first choice where foodstuffs, etc. were involved. Copper and some of its alloys will form a natural patina on their surface. This is an attractive green colour and is a complex oxide/acetate which takes some time to form under normal circumstances.

Many of the Victorian cities of the UK have a green appearance from above, much of which is the result of copper sheeting used for roofing. This was fitted without any treatment and has stood the test of time, and applies to many industrial towns where the copper roofs are still *in situ*.

There are no simple techniques such as chromating or phosphating for producing artificial oxides of copper or its alloys. The investigator, having found a corrosion problem, will have to decide on the best technique available. If it is an aggressive acid and particularly of an oxidizing type, then it would be advisable to change to stainless steel.

The alloys of copper which commonly corrode are those which come under the term of 'brass'. They are prone to de-zincification and would require either replacement with stainless steel or to be painted if they had to continue in operation. They are also prone to season cracking (which would now be called stress corrosion) particularly when components are exposed to ammonia.

Where the red colour which is characteristic of de-zincification against the yellow colour of brass has been found it might be possible to clean the surface and paint it to prevent further damage. There are special paints available and the Copper Development Association in London should be contacted for advice.

Other metals and alloys

Zinc Zinc is a relatively low cost material not normally used for high integrity components. It has reasonably good corrosion resistance, particularly if it has been chromated prior to entering service. The standard chromate passivation on galvanized, electro-plated or zinc sprayed components will give excellent corrosion resistance in the normal environment and acts as a key for painting where this is considered necessary.

The investigator will more commonly encounter zinc failure as a corrosion protection on steel than on components manufactured from zinc.

Nickel and its alloys Nickel has excellent inherent corrosion resistance, particularly under oxidizing conditions and is chosen where there is a hot oxidation environment or where certain types of aggressive corrosion exist. The most common outcome of failure investigation will be as the result of sulphur/sulphide attack.

Under some limited circumstances of heat cycling there could be oxide penetration from the surface resulting in problems.

There is a sophisticated and quite expensive process of aluminizing which can improve the corrosion resistance of nickel under both sulphide and oxidation conditions. This process is analogous to the aluminizing of iron and steel, but is carried out under much more controlled conditions. Normally the process requires rather sophisticated equipment and specialized knowledge; this specialist advice should be sought by the investigator.

Titanium and its alloys Titanium metal without doubt is the most corrosion-resistant material used in any quantity by industry. This applies only to ambient and temperatures up to approximately 300°C maximum. Above these temperatures titanium will form oxides, nitrides, hydrides or a combination with almost any other element with which it comes in contact. The investigator finding corrosion problems from this temperature effect has limited choice in overcoming this as such temperatures.

Titanium is only used on relatively sophisticated equipment and is a high cost material. Thus it is not really possible in a general book of this type to give specific advice where corrosion has occurred.

Titanium can be anodized in exactly the same way as aluminium, that is by being made the anode in an oxidizing atmosphere with a direct electric current. This produces an adherent coherent oxide film which has a better ability to withstand temperature by a significant degree but obviously once damaged will cause considerable problems by being a source for local corrosion and thus a stress raiser.

Specialist advice would obviously be necessary where any titanium corrosion existed.

Cobalt and its alloys Cobalt is almost invariably chosen because of its specific corrosion resistance to the environment involved.

As far as the author is concerned, there are no techniques available which would significantly improve the corrosion resistance of cobalt or its alloys.

Magnesium and its alloys Magnesium and its alloys are extremely prone to oxidation. They can react to ignite spontaneously as powder or thin film form. This obviously considerably restricts the use of magnesium in industry.

There are however, treatments of chromating and anodizing which, provided the surface is not damaged, will give excellent corrosion resistance. The chromate produces a black oxide film which is very thin but makes an excellent key for paint. The paint must be applied carefully and immediately after chromating to ensure good adhesion. Great care must be taken so that all surfaces are covered firstly with a primer and then with adequate top coats.

Magnesium can also be anodized where the same comments would apply as above. The anodic film is better than the chromate film but again would be readily damaged.

Tin and its alloys Tin is little used in modern industry. At one time it was used extensively for tin plate. Where the tin is applied as a very thin film it was found to be slightly porous, thus it was not the tin which corroded but the substrate, normally mild steel. Most modern tin components containing food have a lacquer applied on top of the tin, to obviate any problems with tin porosity.

While tin is non-toxic, many tin compounds are quite poisonous and any investigator finding corrosion on tin plate in contact with foodstuffs must find the reason for the attack on tin.

An alternative coating material such as aluminium or stainless steel would be preferred for aggressive foodstuffs.

Lead There is no normal circumstance where lead could be expected to corrode and would require protection. Lead is almost the ultimate in low-cost materials for use in aggressive environments. Lead has equal corrosion resistance under most circumstances to stainless steel. Titanium would be a superior high-cost alternative.

Tungsten There are no known circumstances under which tungsten could be found to be corroding.

Contamination/corrosion failures involving tungsten are not of the tungsten itself but of the matrix holding the tungsten or tungsten carbide in position.

It is not considered necessary or feasible to apply corrosion protection to tungsten or its alloys.

Non-metals and compounds

Stone, brick, etc. Corrosion/contamination/degradation of stone, cement, etc. will normally be the result of environmental attack. There is an ongoing discussion at present on the techniques which are available and preferable to overcome this problem. These are basically:

- Blasting
- Chemical treatment
- Pressure washing

All of the above have their adherents and opponents. There is information now that some forms of chemical treatment result in more damage than existed prior to the treatment! Blasting and water washing can remove the corroded surface and with modern city environmental control damage is less severe.

Where concrete is involved, the contamination product will often be based on carbonates, some of which do little damage but others may cause problems. Whether to merely wash and scrub or to have some form of mild chemical treatment to neutralise the carbonate is an area where specialist advice is advisable.

Where brick/stone/cement, etc. require painting, this primarily is an area where 'taste' becomes involved and there are many who are reluctant to preserve these types of surfaces at all, preferring the natural look.

Where painting has been applied to a stone/cement surface, very often this causes spalling. The faulty surface should be prepared by mechanical means, that is light blasting or pressure washing. Where repainting is required, specialist advice will be required depending on the type of stone, amount of paint left, etc. There are associations such as the National Heritage, National Trust, etc. where advice can be obtained if old buildings are involved.

Glass/ceramics These in general will not be subject to corrosion or degradation but may be affected by contamination. Attack on glass or ceramics will require extremely active acids, hydrofluoric being the most usual. This is an extremely unusual acid and will not be available except under very special conditions, which are normally controlled.

Contamination of glass/ceramics could occur and would then require specialist advice regarding removal. Glass can be more readily scratched than is normally appreciated and emery paper and suchlike materials should not be applied.

Ceramics and other refractories may have constituents which are susceptible to attack by solvents and care must be taken that trials are carried out before treating surfaces.

Wood With very few exceptions, all wood products will require surface protection before entering service. The exceptions would be some of the sophisticated hard woods where the natural surface will be non-porous and very attractive.

Other hard woods, however, are normally polished and occasionally are still French polished, although more usually lacquer/varnish would be applied.

Soft woods are quite porous and unless sealed, contamination or degradation will take place. Once the wood has absorbed moisture, the rot will take hold. It might be visible as a surface fungus but probably will be progressing inwards unseen.

Treatment would involve removal of all contaminated material and treated to ensure that the degradation will not continue or recur. The surface would then be sealed with a quality primer, many of which are 'active', to prevent further fungal attack.

This is an area where specialist advice is essential, particularly at the initial stage where a long life is required.

Soft woods, once the paint is damaged, will degrade quite rapidly. One advantage of hard woods is that even when the surface is damaged, the degradation is much slower. In such cases degradation will be unsightly but will not cause extensive structural damage.

Normally outdoor wood will require treatment at regular intervals which may be as frequent as 2–5 years but may be longer depending on the type of wood, treatment involved and the location.

There are various associations where specialist advice can be obtained.

Plastics/polymers These are often attacked but whether the term 'corrosion' can be used is a matter for debate. Many organic solvents commonly used in households such as paint thinners and nail varnish removers, will significantly attack plastics/polymers.

The investigator, having seen the evidence of attack, will have to decide whether the material is continuing to be degraded or whether the surface is stable.

In most instances there will be a visible effect but in industry there could be problems of contamination where sophisticated materials are involved, in particular where composites exist, and where problems may proceed without recognition. This is an extremely complex area and it is not possible to give general advice. Components such as a printed circuit assembly would come under this heading. This can have a relatively cheap simple composite plastics/polymer board consisting of glass fibres and plastics/polymers. This may then have a lacquer, and tin–lead tracks, some of which might be gold plated. The components involved can be in ceramic, plastic, glass or metal packages and certainly will have metal connectors. These will have been soldered and the solder will have been applied with a flux which may or may not have been active. The resultant corrosion/contamination/degradation can be extremely complex, requiring individual examination and advice.

In the author's experience, most problems with plastic/polymer components, including composites where they are involved, are the result of misuse – either by leaving them in strong sunlight which results in degradation of the surface, or by the misuse of solvents for cleaning purposes. Very seldom will it be possible to reverse the process which caused the problem and thus specialist advice will almost invariably be necessary.

One problem with plastics/polymers is that the degradation process can continue over a period of time and the end product may be extremely brittle. In the case of plastic boats, for example, degradation can be so advanced that a very light impact can cause serious problems to the hull.

Natural materials – wool, leather These do not degrade at the same rate as artificial equivalents under sunlight conditions. The dyeing process commonly used with wool, however, can affect the structure, resulting in brittleness, and this process cannot be reversed.

Contamination of wool and leather will normally be solved by washing with water and mild detergent. It is not normally advisable to use solvents on these natural materials as they have natural oils of various types as part of their make-up, and solvents will remove these oils from their surface.

ABRASION, EROSION, GALLING, WEAR

A material harder than another will cut the softer material but a soft material will not cut a harder material. Materials of equal hardness can, however, cut each other.

In the Bronze Age a lubricated bronze wire was used as a saw to quarry marble. Powdered marble or sand was fed on to the moving wire until a slot was produced, whereupon the marble debris continued the cutting action. Modern quarrying still uses this technique, with continuous power drive of the cutting wire.

It is possible for pressure to cause swaging, that is movement of metal. With two metals in contact, both materials will swage when pressure is applied. Even soft steel can swage a harder steel if sufficient pressure is applied. Swaging is a compression-type mechanism, but can be difficult to differentiate visually from abrasion, as it commonly results in abrasion either as the surface metal breaks off, or as work hardens and cuts.

The mechanism of abrasion is identified by the loss of material or presence of loose particles, and can be a reason and mechanism of failure.

Where abrasion results from particles in a liquid flow, there will be a characteristic appearance of the defect. The investigator will recognize this by the surface appearance – as if a horse had been walking upstream!

The first diagnosis and visual examination will have identified that the problem is abrasion, erosion, wear or galling. In these cases it is possible that the initial indications are noise, vibration or fluid leakage. But ultimately it is possible that abrasion or wear has resulted in seizure, fracture or fatigue initiation.

The visual examination should confirm that any fatigue or tensile failure was initiated at a point where wear or abrasion took place. As with corrosion it is desirable that the investigator is involved at an early stage and is able to collect debris from the site for further investigation. The abrasion could be from foreign bodies such as sand or glass which will cut the surface and could be ground to a fine powder. The evidence for this is essential in determining the reason for the failure.

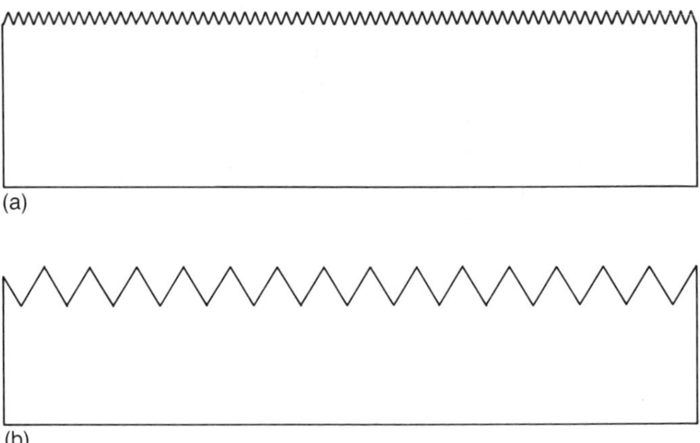

Figure 7.24 Examples of surface finish: (a) 45 peaks per unit length; (b) 10 peaks. The same load applied to (b) would create 4.5 times the load applied to each peak in (a), causing possible fracture and the means of abrasion damage due to build-up of abrasive particles.

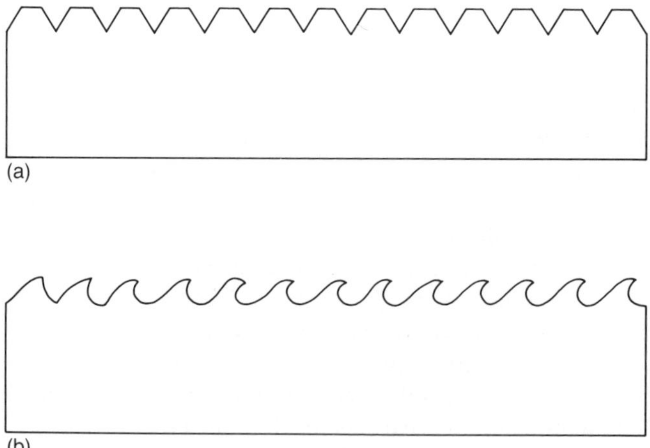

Figure 7.25 Abrasion damage can commence because of a rough surface either when (a) the peaks are worn or sheared off, or (b) the peaks are smeared or compressed before breaking off.

With non-metals abrasion will result in the same visual effects, but it is not normal for non-metals to work-harden. Break-up of non-metals will not be accelerated, as the particles so produced will not increase in hardness.

When the same load is applied to a smooth surface as to a rough surface, it will be seen from Figure 7.24 that the load will be much higher on the peaks of the poor finish, which can result in the peaks bending, wearing or fracturing. Where the peaks fail, particles are produced which will score or abrade as in Figure 7.25.

Where bending of the peaks occurs there will be an increase in clearance, and this too can result in problems.

The visual examination in most cases can differentiate between swaging – where the surface peaks are distorted, and abrasion – where the surface breaks up. The visual effects of abrasion, galling and corrosion damage, however, can be confused by heating effects (from lubrication problems).

Cavitation

Cavitation erosion is a very specific form of metal removal which is quite different from corrosion and other forms of abrasion. The mechanism of cavitation erosion is that the surface of the metal is subjected to liquid pressure which rapidly becomes to a tensile force. That is, at one instant the surface is being compressed and a very short time after it is subjected to tensile stress. This was first identified in ships' propellers, where at a certain point on the revolution the surface of a blade is pushing the water, but a critical instant later the water is sucking at the surface of the same blade.

Where liquids of low vapour pressure are subjected to this push/pull cycle, gas bubbles form in the liquid at the metal surface. At the end of a stroke the gas returns to a liquid of a much smaller volume, thus creating the tensile load on the metal surface.

Liquids such as water probably allow cavitation to occur even when remaining liquid as the compression load at the surface rapidly changes to a tensile load. This effect can be demonstrated by cupping both hands under water and closing and opening the palms. The palms will rapidly become very painful as the surfaces are repeatedly pulled apart.

Areas where cavitation erosion has been identified include thin walled cylinders which are liquid-cooled in combustion engines which can vibrate and produce compression and tensile forces. See Figure 7.26.

Reciprocating pumps and propellers operating in liquids are the more common sites for cavitation, which is readily recognized by the experienced investigator by the presence of circular pits, as in Figure 7.27.

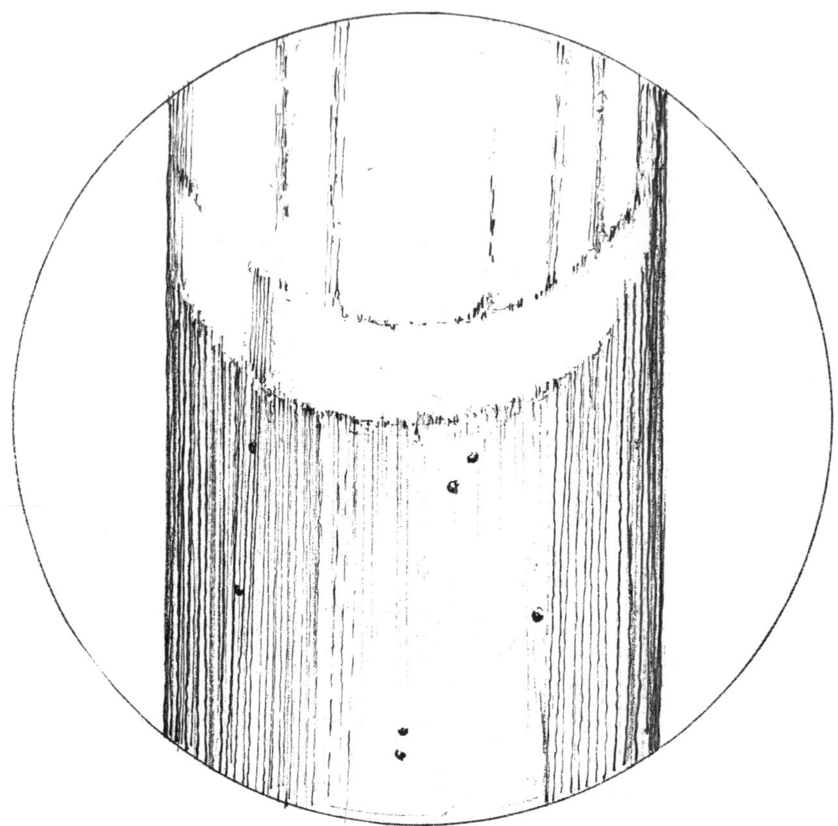

Figure 7.26 A chromium plated reciprocating shaft subjected to vibration within a liquid. Failure resulted from cavitation forces plucking grains of chromium from the surface, which then abraded the shaft.

These are well defined and are quite different from the pits produced by erosion, which will always have a channelling effect, and corrosion, where the pitting is rough.

Cavitation can be cured by engineering design, and there are now alloys which are resistant to cavitation. Cavitation is not a mechanism that could be expected to be identified on any non-metal.

THERMAL FAILURE: EFFECTS OF HEATING/COOLING

When materials are either heated above or cooled well below ambient conditions problems of a thermal design nature can arise. For example,

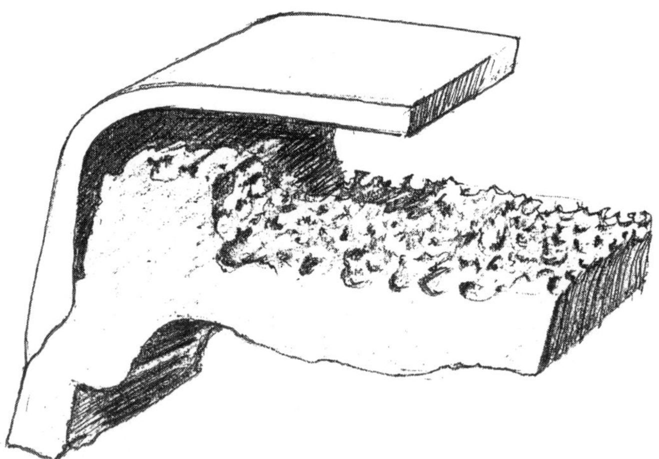

Figure 7.27 Cavitation erosion of large mild steel brake drum. The under surface is the friction face; the inner surface is water-cooled, with an annular stainless steel shroud. During braking, there is severe vibration resulting in cavitation forces eroding the mild steel with no effect on the stainless steel. The illustration shows a 30 mm section cut from a 4 m diameter brake drum.

when metals are heated their tensile and yield strengths can reduce. The ductility is also increased and impact properties will probably improve to some extent.

When metals are cooled the tensile and yield strengths will usually increase, ductility will decrease, but impact properties can worsen dramatically. These effects will all be found by the visual plus micro-examination with mechanical testing sometimes required.

Visual effects seen from over-heating could be the colour effect on the surface of metals. Depending on the surface finish, the time and to some extent the atmosphere, steel will discolour in a predictable manner with the colours giving some indication of the peak surface temperature reached. These effects were used by the traditional blacksmith for tempering with colours like 'straw', 'blue', etc. giving him an idea of the temperature of the component.

For interest, the following table gives the temperatures and colours which are found with steel. It must be appreciated that the colours will vary depending on the heat time, surface finish, and to some extent the atmosphere. Where however, the investigator sees these colours on steel components, he will be aware that a temperature effect has existed. If this temperature effect is above the tempering temperature then the steel can be expected to soften.

Table 7.4 Discoloration of steel due to temperature

Temperature	*Surface colour*	
220–230°C	Pale yellow	'Straw'
240°C	Dark yellow	
255°C	Yellow/brown	
265°C	Brown/red	
275°C	Purple	'Blue'
285°C	Violet	
295°C	Light blue	
315°C	Dark blue	

To estimate the temperature with any degree of accuracy, consistent surface finish, atmospheric control, time at temperature and type of heating must be controlled. These temper colours must, therefore, be treated with caution. They can however, give a reasonable indication of the temperature at which the component has been held.

If the temperature is high enough then the surface metal will combine with the atmosphere. If this is a normal atmosphere then it will oxidise; if there is any contamination in the atmosphere then the oxidation may be accelerated and other compounds can be formed as discussed under Chemical Corrosion. The corrosive products will normally be abrasive and if readily removed from the surface could cause abrasion in moving parts. These effects will all have been identified and hopefully understood at the first diagnosis and visual examination.

Micro-examination would then be carried out, firstly on a portion of the component which had not been affected by the heat, and secondly on the portion which had been heat affected. In the case of steel the metallurgist would obtain a very clear indication of any effect of heat on the structure. Chapter 14 describes the heat treatment of steel in layman's terms, and what the metallurgist can expect to conclude.

Where there are rapid changes in sectional thickness, thermal expansion/contraction can result in stressing or cracking, which might then initiate fatigue. If high or low temperatures are rapidly applied, there can be cracking due to the rapid thermal cycling. Figure 7.28 shows a typical example of thermal failure.

In order to produce cracks from heat alone, relatively high temperatures, such as are found at welding and in gas turbine engine components, would normally be required.

The phenomenon known as 'creep' is the result of long-term tensile loading at temperature. It can be identified visually with some metals such as lead and tin, and also with plastics. This will be seen as a drooping or stretching phenomena. There will, however, be no visual evidence to observe of

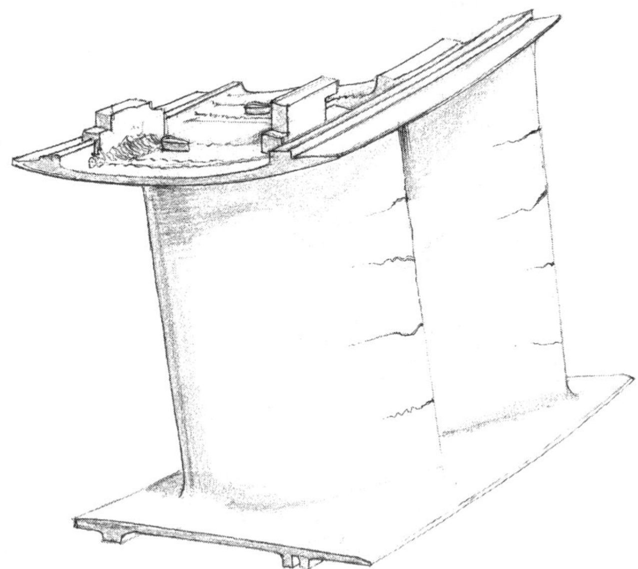

Figure 7.28 Typical thermal cracking on the thin (trailing) edge of a gas turbine nozzle guide vane.

creep on many components. It requires an experienced metallurgist to identify creep voids after careful metallurgical preparation. Thus creep, like corrosion, can be suspected as a failure mechanism (the How), but it requires further skilled investigation to identify the reason (the Why) for failure.

The above information applies to metals. Other materials such as plastics/polymers, wood, concrete etc. are quite different. They can 'age', that is their properties differ with time. Almost invariably increase in temperature will accelerate the ageing process. The presence of ultra-violet or infra-red light can have a significant effect on the ageing of plastics/polymers. The effect will generally be to make the material more brittle. Some plastics, however, become very soft and ductile over a very small temperature rise.

Evidence of creep in some plastics/polymers is readily seen, with a cold water pipe retaining a straight horizontal line, while the same plastic/polymer warm water pipe will droop between supports.

Other plastics/polymers under identical conditions will remain rigid but can experience the hot water pipes becoming very brittle, thus failing under low impact, or even expansion/contraction loading.

Concrete, unless specially formulated, will not reach its specified properties if subjected to frost damage during the period after pouring.

Grinding abuse, stray electric arcing (stray welds), welding problems in general, could all be classified as thermally-generated problems. Grinding abuse is where the grinding wheel in contact with the metal being ground raises the surface temperature by friction above the critical temperature, either softening, hardening or, under some circumstances, cracking the material. The cracking can be the result of a very thin layer of steel being quench-hardened, then shrinking as it cooled – in essence quench cracks.

PART THREE

Failure Investigation

8 Micro-examination

After visual examination, micro-examination is the most important aspect of failure investigation and for failures in metal components is the reason a metallurgist must be involved. Engineers, physicists, joiners, accountants and lawyers can all be trained to carry out visual examination which will identify important features but it requires the skilled metallurgist to choose, prepare and interpret specimens from critical areas.

Correct micro-examination requires firstly the professional skill to identify the areas where specimens should be cut. Secondly, there is the technical skill to correctly cut, mount, polish and etch the chosen specimens. Finally there must be the metallurgical ability to recognize the structure and effects produced.

PREPARATION FOR MICROSCOPIC EXAMINATION

The qualified metallurgist must have the ability to recognize the different structures and it must be appreciated that examining a structure at ×100, ×500, ×1000 and higher magnifications can give different stories. It is seldom necessary for a failure investigator to examine a specimen at higher magnifications than ×1000. Any metallurgical structure not identified at ×500 or a maximum of ×1000 will be of so little significance that it will not result in failure. Any problem only found at these higher magnifications would mean that a sophisticated design problem exists, with an unacceptably low safety factor.

Most failures occur at stress levels well below the design requirement of the component or assembly. If the design is such that 100 units of force will be required before failure (however identified), then most problems occur at the 10 to 25% level. Aircraft-type components however are expected to operate at up to 75% of the design force. If the design requires 95–99 units to be applied before failure then the metallurgist will require to examine at high magnification. In other words, most components have a 'safety factor' of

75–90%, and aircraft type pieces 25%. Only parts designed with safety factors of 1–5% would require high magnification examination because the investigator is looking for small amounts of degradation in a product operating close to its design limit. These very low safety factors would not normally be acceptable economically, or ethically.

High magnifications are commonly required for research and development, but not usually for failure investigation. Degradation and failure are normally well advanced with products operating with high safety factors.

The metallurgical microscope is different from the binocular microscope used for macro-examination and in biology, botany, etc. The latter makes use of light fed to the side or below the stage; the light is then thrown up across or through the specimen, which is examined from the top. Full use is made of light and shade and 3-D effects.

Metallurgical specimens are solid, do not transmit light and because of their polished finish, reflect light. It is thus necessary to shine the light through a prism above onto the specimen then view the reflected light through the lens to the eyepiece. This means that more intense sources of light are required, and the light train from the source to the eyepiece must be very carefully aligned and maintained. The visual effect exhibits no shadows and produces a 2-D image which is often in full colour not visible with binocular microscopes.

The use of etches

The specimen examined in the unetched condition, will show no metallurgical structure but cracks, non-metallic inclusions or gross differences in hardness of the surface will be visible. It is essential that all specimens are first examined in the unetched condition. This will firstly show whether or not the standard of preparation is acceptable as well as the non-metallic inclusions, cracking, porosity, etc. which when very small are difficult to see after etching, but could be significant.

Etching using various chemicals reveals the metallurgical structure. What happens is that the chemicals attack the different constituents of the metal at different rates and thus there will be some areas which have had more metal removed than others. The light reflected from the different areas will thus be different in profile, texture and colour.

There are now a host of etches which give consistent results regarding the degree and effect of attack and always reproduce well. The metallurgist is thus able to recognize the different structures from the etched specimen. The technique of mounting, polishing and etching is a most important part of the metallurgical examination.

The area chosen for examination will have been decided at the visual and macro-examination. It is good practice to take a further specimen remote from the point of failure, as this can be used to confirm the base metal and which should be representative of the material used to produce the faulty component.

The selection of the specimen

In choosing micro-specimens from close to, or at, the point of failure, if both faces of the fracture have been submitted, then the metallurgist can quite happily destroy one fracture face to produce various specimens which will identify the structure or cracks at the failure initiation. If a fatigue has been identified, then the site of fatigue initiation may be clear but it will be necessary to identify whether there is any metallurgical defect present. If the metallurgical defect is of a linear nature then the direction will be critical. It must be appreciated that the micro-examination is on a single plane and the metallurgist is examining a small surface area, and any crack-like defect at right angles to this surface will show in the cross-section. This is why, for example, slag stringers 25 microns in diameter, and 100 microns long which exist at the surface and act as the stress raiser, can be difficult to find.

If a linear defect exists and the direction is not known, then the metallurgist is in a quandary regarding where to cut the micro section. A section cut in the same direction, that is parallel to a linear defect, could completely miss the defect or destroy it at polishing (see figure 8.1).

If the specimen is cut at right angles to the defect then the defect will be visible as a cross-section. It will be appreciated therefore, that if only one fracture face remains, skill or luck is required in how to cut the specimen. With both fractures available, if nothing appears in the first micro-examination, the other fracture face can be cut at right angles to the first cut.

Cracks identified at the visual examination and not opened are cut at right angles to give a cross-section. Opened cracks will be cut at right angles to the crack surface.

Any welding will have a cross-section taken of the weld, and as welds are often non-homogeneous, more than one micro-section will invariably be prepared.

The selection of the micro-sections therefore requires very considerable care and only experience and training ensures that this is carried out correctly. It is the skilled visual examination which ensures that micro-specimens are cut from the correct area. No amount of technical or professional skill will identify a metallurgical defect if the micro-specimen is taken from the wrong area. The practice of random cutting of specimens without careful visual examination is unfortunately quite common.

Examination of the initial micro-specimens prepared to show the situation remote from the fracture and adjacent to or at the fracture will generally indicate whether there is any peculiarity in the structure which might require further examination.

In most failure investigations a correctly carried out visual examination followed by a macro-examination will give sufficient information to the experienced metallurgist on where to cut and prepare the sections to give the evidence

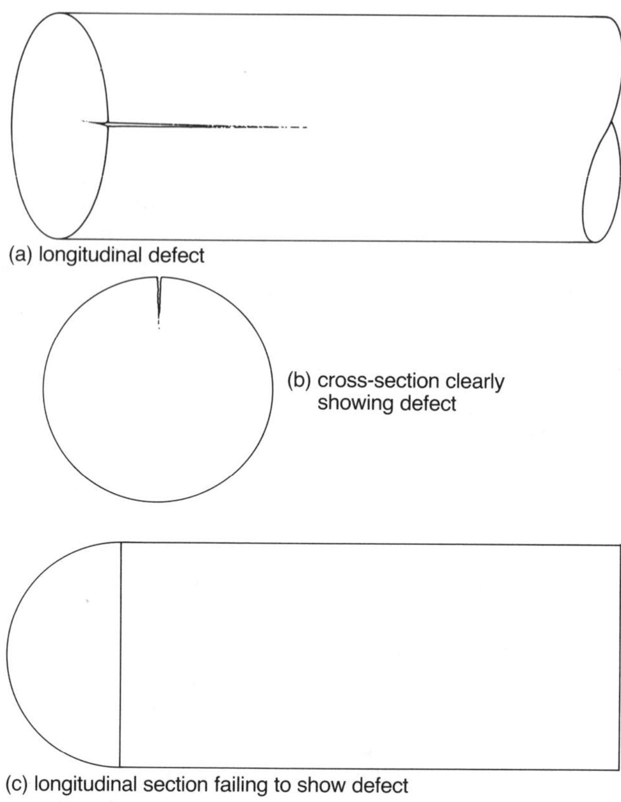

Figure 8.1 Crack in bar with longitudinal defect which when the bar is sectioned could easily be missed unless the section is across the bar.

required to confirm the visual and macro-examinations, using two or three micro-specimens. The investigator, however, must always be prepared to cut and prepare other specimens if the initial examination shows structures which are not consistent.

Once the choice of specimen is made, it must be carefully prepared to make certain that there is no distortion or any spurious effect such as over-heating caused by faulty preparation. If flame cutting is used, this must be carefully carried out and must not be within 50 mm of any critical surface with the flame-cut components kept cool to ensure that heat transfer does not affect the metallurgical structure.

Cutting with rotary slitting or grinding also requires care and cooling to ensure that no over-heating takes place. In general the ideal cutting method is by hacksaw if this is possible. The specimens cut by hacksaw will be rough and should be linished with coolant to prevent local heating.

After separation or cutting, the specimen(s) should be examined to make certain they are not overheated. This would be shown by colour on the cut faces such as 'blueing' or a yellow colour. Any such indications must be treated with suspicion.

Mounting and polishing the specimen

Mounting is necessary to secure the specimen and produce the flat surface essential for micro-examination. There are various techniques used, generally based on thermosetting plastics. Some are cold cured, others hot sometimes with pressure applied, but it is essential that the plastic should expand at the setting operation to make a tight and secure bond between the surface of the plastic and the specimen.

Polishing may produce a radius at the edge due to the specimen surface showing defects. The more modern technique of plastic pressure moulding overcomes these problems giving an excellent bond to the specimen ensuring no leakage of etchant, a sharp and well-supported edge, and no reaction of the plastic with the etch. Some plastics used in mounting specimens are clear, and it will be found that individual metallurgists have 'fads' regarding size of specimen, type of mould and plastic used.

The specimen size will very often be determined by the metallurgist deciding how much he wants to see. Most moulds are circular, about 25 mm in diameter, thus restricting the size of the micro-specimen. With modern polishing techniques larger specimens can be polished but many metallurgists would rather prepare two adjacent specimens, rather than one large one. The specimens thus are very seldom more than 20 mm × 20 mm, and often much smaller.

Great care is required in placing, and if necessary supporting, the specimen prior to mounting in plastic to ensure that it is the correct face that is polished. It is now common to use clear mounting plastic which allows the operator to see how the specimen lies in the mount, but it should be appreciated that only the operator producing the specimen will really know for sure which face is polished. Operators must be instructed to remember or record the way the specimen is mounted as it can be embarrassing to find that the crack readily seen at visual examination does not appear at micro-examination because the specimen has been mounted at the wrong angle, or has fallen over during mounting or curing.

The polishing technique also requires considerable skill and to some extent is an art or craft. Modern equipment uses rotating tables with a flat surface on which is held the abrasive disc, generally paper which is replaced once it is contaminated or torn. There are now fully-automated polishing techniques available but these are seldom used by investigative metallurgists, who prefer the flexibility of traditional methods of hand polishing.

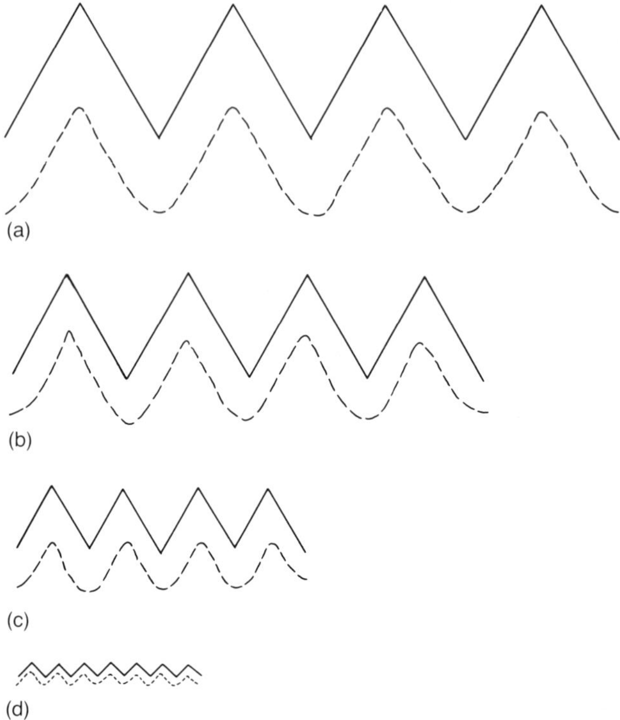

Figure 8.2 Diagrammatic representation of stages in correct surface preparation of micro-specimens. (a) Rough polish stage with peaks sharp and evenly spaced. The lower line represents the limit of the distorted layer. (b)–(d) The peak dimension diminishing with the distorted layer reducing after successive stages of polishing.

There are broadly eight stages in the polishing process between cutting the specimen and microscopic examination. These are:

1. Rough polish to remove sharp edges and produce a flat surface
2. Mount the specimen in plastic
3. Polish on 120–220 mesh (125 microns)
4. Polish on 300–500 mesh (40 microns)
5. Polish on 700 mesh (30 microns)
6. Polish on 1000 mesh (10 microns)
7. Diamond polish 6–8 microns
8. Diamond polish 2–4 microns

For diagrammatic representation see Figure 8.2. At each stage the specimen is held in contact with the rotating table or moved by hand back and forth on static flat plates. The specimen is moved in a narrow figure-of-eight until all the surface abrasion marks or scores are roughly in one direction.

The specimen is then washed under running water, turned at right angles and the process repeated for each stage. All evidence of surface scoring from the previous stage must be eliminated before moving to the next stage. The aim of polishing is that the surface must be abraded to a fine finish and not smeared or cold worked. The surface below each cut will have a distorted layer at approximately the same depth as the height of the scratches. Thus as the scores or scratches become finer, the depth of the distorted layer is reduced. It is generally agreed that it is desirable, if not essential, that those preparing micro-specimens, and the qualified metallurgist, should have detailed knowledge of the techniques used and problems which can exist. It is very easy to physically distort the surface of a micro-specimen, giving results under the microscope far removed from the true structure which can give meaningless or misleading information.

Pre-etch examination

The washed and dried specimen is examined with the metallurgical microscope. This examination is first performed at low power, approximately ×20–30 magnification, with the surface examined for any remaining scratches, scores, etc., and any evidence of gross non-metallic inclusions, pitting or cracking. This will show the metallurgist or technician if the standard of preparation is satisfactory, or requires repolishing.

The specimen is then examined at approximately ×100–150 magnification. Particular care at this magnification must be made of the specimen edge looking for any surface cracking, oxidation or evidence of other imperfections with any peculiarity noted.

If the surface has been prepared well with a minimum of scratching or smearing, the magnification is raised to ×400–500 to continue the examination. Normally, this will be in detail at the edge of the specimen with an additional scan across other areas, particularly where any evidence of peculiarity has been noted at the lower magnifications. This examination is intended to identify imperfections, cracks or porosity, and notes should be carefully made of all observations.

If necessary, the specimen should then be examined at ×1000 but in most investigations this is not necessary.

ETCHES AND ETCHING

The specimen is then etched. A brief list of etchants and why they are chosen are given in Table 8.1.

Table 8.1 Types of metallurgical etch

Etchant	*Mix*	*Uses*
Nital	1–3% nitric acid/alcohol	Mild and alloy steels
Ferric chloride	5% ferric chloride in 1–5% hydrochloric acid	Stainless steels; nickel alloys; some copper alloys. Used to identify creep in voids in nickel/chrome alloys
Marble etch	10% copper sulphate in hydrochloric acid	Stainless steels; nickel alloys
Sodium hydroxide	1–5% NaOH (caustic soda) in water	Aluminium alloys
Ammonium persulphate	10% persulphate in 10% ammonia. Make up fresh each time	Aluminium and its alloys; copper and its alloys
Picral	10% picric acid, 10%	Sigma phase in stainless steel: hydrochloric acid in alcohol alternative to ferric chloride
Aqua Regia	Concentrated nitric acid and hydrochloric acid in various ratios	Some stainless steels; titanium and its alloys;

There are many other specialist etchants but the above will generally satisfy the investigative metallurgist.

The etchant is designed to attack the different metallurgical structures at different rates, removing more metal from the more active constituent than from the less active one.

Great care is required in etching with immediate washing and drying to ensure that little or no staining occurs. The more active the etchant, the more readily it will stain. Practical metallurgists make use of the swabbing technique, where the specimen is warmed under running hot water, and then gently swabbed with cotton-wool soaked in the etchant. As soon as the etched structure appears the specimen is returned to the hot running water. The specimen must be immediately removed from the hot water and either blow dried with hot air, or immersed in acetone. This is essential to prevent further staining.

There are now available 'mass production' techniques for etching, including electropolishing, which are useful in process monitoring. These are seldom of value to the investigative metallurgist who requires flexibility rather than the more rigid high productivity techniques.

It is preferable to under-etch as over-etching can hide some effects and it is easier to repeat the etch than it is to repolish.

Post-etch examination

The specimen is returned to the microscope and examined as before at low magnification of ×20–30, medium magnification of ×100–150 and then a magnification of ×500 and, if necessary, ×1000.

The structure is examined, the effects identified and noted and the specimen then returned to the technician for reworking. Exactly the same polishing technique as before is used. That is, the specimen is dried after polishing and re-examined. This examination is to remove the effects of etching and make certain that any inclusions, cracks, surface effects, etc. observed after etching are still present.

Any difference in the effects noted must be recorded, and the reason identified. This will generally mean re-polishing on at least the final paper plus diamond polishing. Once the metallurgist is satisfied that consistent results have been obtained, or that any inconsistency is the result of variation in the specimen itself and not a polishing/etching effect, then the results seen can be logged as a permanent and true record.

In many laboratories the responsibility to produce the specimen and make sure that the second examination shows exactly the same results as the first is that of the technician, who will probably be a junior or trainee metallurgist. If there is any peculiarity at all then it is essential that additional examinations are carried out. There is a history of etching effects being reported as metallurgical structures, as the slightest stain caused by delays in etching, or peculiarities in the etching solution, can result in very interesting structures. These have been recorded and reported, but could not be reproduced. These are real issues of practical metallurgy which cannot be ignored.

RECOGNIZING METALLURGICAL FEATURES

Where the failure is a fracture, or a single crack which has been opened, the metallurgist will prepare a specimen to show the surface. The quandary of how to cut the specimen if only one fracture face is presented has been described earlier in this chapter.

The investigator will look for one or other of the effects discussed with cracking. The effect of cold work, excessive grain size and a well-defined brittle metallurgical structure will be easily recognized and their significance appreciated.

Porosity is a common problem with castings and rare with wrought items. The size and distribution of the porosity can be significant as can the type. Clean fine porosity, free of oxide and distributed through the section with no rough edges, will be the result of casting shrinkage. This, while not advantageous, need not result in a serious effect on mechanical properties.

Porosity which is oxide-filled and concentrated at local areas with rough edges will normally indicate faulty casting control, which can be expected to lack ductility and to be brittle. Surface porosity can act as a fatigue stress raiser. Porosity identified at micro-examination will commonly result in macro-examination to quantify the problem.

It is the surface of the component associated with the failure initiation that will be of most interest to the investigator.

The reason why the crack or fracture started at a particular point must be identified if possible together with information on crack propagation.

There are numerous metallurgical effects which the investigator must be able first to see, then to recognize. These will almost invariably be seen and recognized at magnifications of less than ×500, but on occasion some detail may be recognized at ×1000.

Depending on the material involved and the operating conditions, the investigator will have ideas regarding the possible cause. Different etchants would then be used to identify Sigma phase, creep voids, carbide precipitation, phosphide, etc. This is the area where the ancient metallurgist is essential to pass on knowledge which will never be available from book learning. There are a few excellent books containing thousands of microphotographs, but they never seem to have photographed and described the feature presented to the investigator!

Crack identification

Where cracking, surface imperfections and/or porosity are identified on the unetched specimen these should be closely examined after etching to see any evidence of decarburization, or other local surface effects. Cracking should be closely examined at various magnifications. The crack characteristics will supply considerable information on the reason for its occurrence.

Raw material cracks These can be of various types, almost always associated with oxide, and may show decarburization. They are normally readily recognized.

'Weld decay' cracking These will be on austenitic stainless steel and will be associated with the grain boundaries. They need not have been involved with welding, but will have become heated in the 300–600°C temperature range.

Fatigue cracking The investigator will normally be more interested in what initiated the fatigue than the crack itself. This is commonly a tensile burst, for example the root of a stud thread which has been over-tightened. Fatigue can propagate from surface material cracks, corrosion pits, stress corrosion cracks, grinding cracks, sharp radii, surface damage, slag or oxide stringers. All these can be seen firstly in the unetched condition, and confirmed after etching.

Metallurgical defects such as too sharp a demarcation between a case hardening and the more ductile core; large surface grains causing brittleness; grain boundary effects such as temper brittleness; decarburization; and too thin a case can all be identified after etching.

Stress-corrosion cracking These are unique cracks which bifurcate or branch. The metallurgist can estimate whether stress or corrosion is the major contributor to the problem. Where there are a few cracks propagating across the section then stress is the major problem. Where there are a myriad of short cracks, all bifurcating, then corrosion can be blamed.

There is a case history where stress-corrosion cracking of a minute nature was identified which then propagated by fatigue to cause major failures. Minute particles of copper were identified at the surface, and discussion revealed that copper-loaded, anti-seize grease had been used in the process.

Stress-corrosion cracking is now the single most common reason for failure. There is some evidence that the full extent has not been appreciated. Many failures are reported as fatigue or brittle tensile stress, where the initiation has in fact been stress-corrosion cracking.

The stress component to initiate cracking need only be quite low. The same low level of stress, however, can be sufficient to propagate by fatigue from the acute stress raiser presented by the initial crack.

Tensile or bursting cracks These will have no metallurgical effect, will be clean and may be 'gaping' at the surface. The investigator will have to eliminate the above causes for cracking before reporting a tensile burst.

Hydrogen embrittlement cracking comes within this area, but there must be a record of atomic nascent hydrogen involvement before hydrogen embrittlement can be blamed. It can be expected that the metallurgical structure of the component together with hardness testing will indicate some degree of brittleness.

Relevant metallurgical features

At this stage the problem may be clearly identified, with no further work being required. It could have been that the material specification called for a case hardened component, whereas the micro-specimens showed a free-machining mild steel, with no evidence of any case.

Care must be taken, however, that any unusual or peculiar metallurgical feature identified has a relevance to the failure. Visual and macro-examination will have identified the mechanism of failure, and the micro-examination to be useful, must add to this and not confuse the situation. There are many metallurgical features which differ from the specification requirement but have no relevance to the failure. This can present a predicament when report writing, as to whether or not 'irrelevant' information should be included. It would generally be inadvisable in these case to make any comment in the conclusion.

The portion of the report covering the micro-examination and metallurgical aspects will generally require a mention of the structure and its significance. This should be as concise as possible, with any explanation or further comment given in the Discussion.

However, it is not advisable for the metallurgist to ignore the issue. The metallurgist carrying out the investigation must be aware that a second opinion might be requested, and the trouble then arises that with any two metallurgists there will always be at least three opinions!

Failure reports, to be useful, should make clear that some of the information observed has no relevance to the failure, it should only be included and referred to in the discussion section. It would be a poor report which had information in the Conclusion of a metallurgical structure which bore no relationship to the problem being investigated, and was not explained.

The need for further tests

The micro-examination will on occasion give information indicating a mechanical problem could exist, which may require impact or tensile testing to be carried out. For example, a large grain structure could indicate low impact properties, surface defects of several types could suggest propagation by fatigue or bending, while a very sharp demarcation between case and core could result in exfoliation, or shear cracking at the interface.

It may also be indicated that chemical analysis is required to clarify some significant metallurgical effect noted. Where corrosion is involved it may be necessary to involve the scanning electron microscope (SEM). This can supply details of the corroded film make-up or grain boundary effects which could help to identify the source of the corrosion.

The need for past experience

The experienced metallurgist can estimate with some accuracy the carbon content of plain carbon steel in the normalized or annealed condition. The grain size will also supply meaningful information.

The different structures of tempered martensite can be recognized and indicate degrees of ductility or brittleness. A structure of coarse needles of

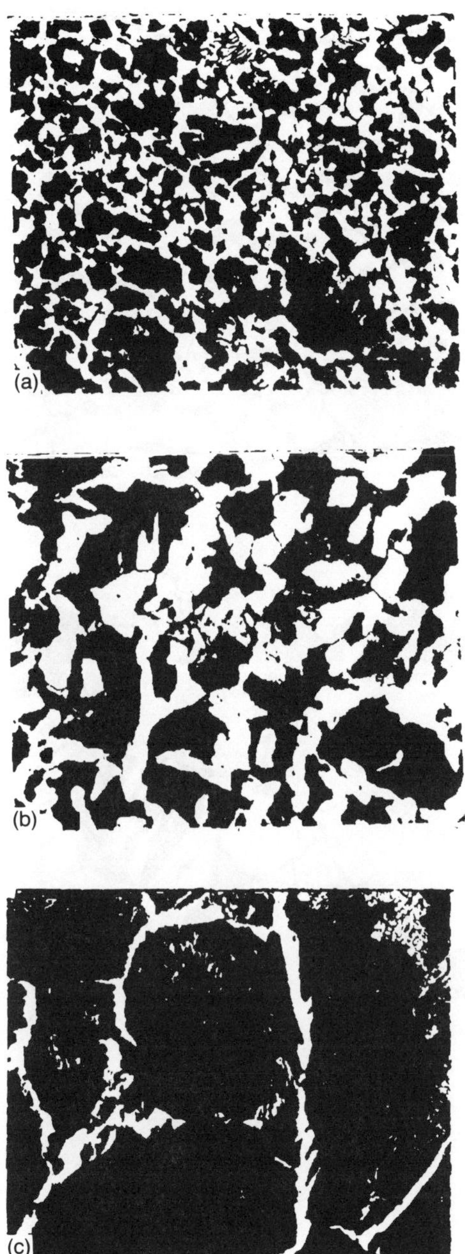

Figure 8.3 Micrographs of 0.45% carbon steel specimens etched in nitric acid/alcohol (nital), at ×500 magnification. (a) correctly normalized; (b) held for too long at the correct temperature; (c) normalized at too high a temperature.

Figure 8.4 Micrographs of 0.25% carbon steel specimens etched in nitric acid/alcohol, at ×500 magnification: (a) correctly normalized; (b) normalized at too high a temperature.

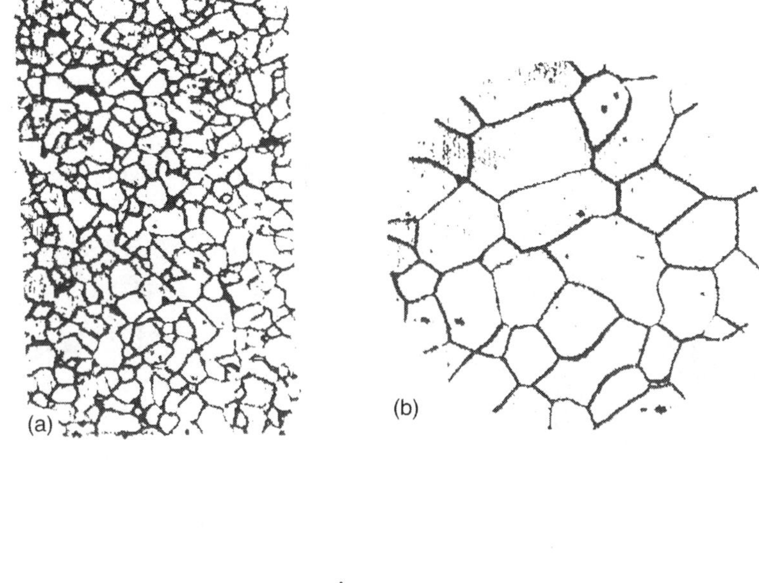

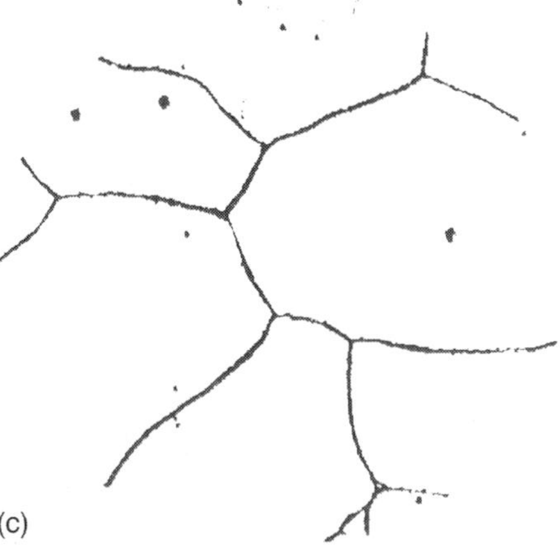

Figure 8.5 Micrographs of 0.05% carbon steel specimens etched in nitric acid/alcohol, at ×500 magnification: (a) correctly normalized condition with fine grains; (b) cooled more slowly, held for a longer time or taken to too high a temperature; (c) overheated, that is held for too long a time at too high a temperature.

martensite would be hard and brittle. A fine grained needle-like martensite would be hard, but less brittle. As the needles become less well defined, the toughness will increase and hardness decrease. The 'gingerbread' structure of tempered martensite is generally the best compromise of tensile strength plus toughness.

With steel the traditional names for the structures range from the very brittle alpha martensite, through beta martensite, primary troostite, sorbite, to the

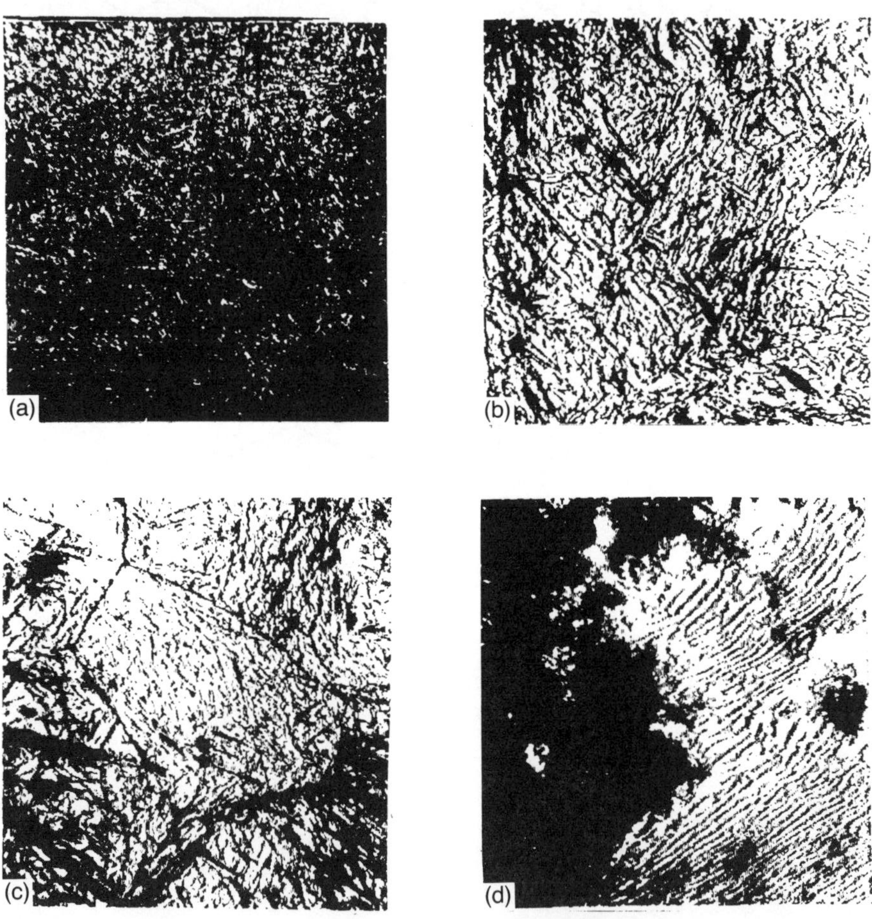

Figure 8.6 Various micrographs of carbon steel at ×1000 magnification. (a) 0.4% carbon alloy steel correctly quenched to give acicular martensite, then tempered at 650° C for 2 hours. This is the sorbitic structure to give the best tensile strength/ductility ratio. (b) 0.6% carbon steel correctly quenched. Structure is needle-acicular martensite with fine needles and no evidence of grain boundaries. (c) 1.0% carbon steel quenched from slightly too high a temperature. Needle-acicular martensite with large grains: steel would be brittle. (d) 0.4% carbon steel which has been correctly quenched. Dark area is troostite (tempered martensite), lamella area is pearlite.

tough bainite or ductile pearlite with plain carbon steels. The modern metallurgist, however, makes less use of these names, commenting rather that the structure is of tempered martensite, then qualifying this with information on the significance of this to the failure.

It is not possible to supply in this or any other book advice which could self-educate a non-metallurgist. There are a host of metallurgical structures which can vary dramatically at different magnifications. There is no substitute for the knowledge of an experienced metallurgist, who must guide and advise the freshly qualified student or investigator.

Cast irons are examples of structures which supply considerably more meaningful information to the metallurgist than either chemical analysis or mechanical testing.

Chemical analysis can only state, for example, 2.5% or 3.0% carbon. The micro-structure will show the distribution of the carbon between graphite, which is carbon, and iron carbide, either as 'white cast iron', or as pearlite. This will dramatically affect the assessment of the mechanical properties based on mechanical testing, and accurately indicated more economically by micro-examination, compared with anything derived from chemical analysis. For examples of micrographs of carbon steel see Figures 8.3 to 8.6. More information is given in Part 4 on Materials.

Non-ferrous metallurgical structures are not always as readily evaluated as ferrous structures. An example is that the optical microscope cannot resolve the structures involved when solution treatment and precipitation hardening are involved.

Under most circumstances this has little relevance as hardness and tensile testing along with the mechanism of failure will identify if there are problems in that area.

Problems such as grain size, grain boundary effects, porosity and non-metallic inclusions, together with surface effects and the type of cracking will be similar or identical to the ferrous specimens.

9 Hardness testing

This subject would normally come within 'Mechanical Testing' as it is a technique which estimates strength. For practical purposes it is non-destructive and can be carried out at low cost on macro/micro-specimens or components, and so is more correctly related to NDT. It is, however, very common that the metallurgical investigation will make use of this test as a routine part of an investigation, and therefore justifies a separate section.

Hardness can accurately assess abrasion resistance of all materials, and with less accuracy the ultimate tensile strength of metals. It can reasonably accurately estimate the ultimate tensile strength of steels, and give an indication of the tensile strength of other metals. In conjunction with visual and micro-examination the results can often identify the reason for failure.

UNDERSTANDING HOW TO GET THE BEST RESULTS

Hardness testing is very flexible, with tests possible in the field with simple equipment, but will normally give a spread of results, so is not an exact science. It is essential that a number of tests are carried out, with 'wild' results being ignored and a minimum of three 'sensible' results accepted – sensible being defined as three results within 10% of a mean.

Before and after any series of tests, it is essential that calibrated test pieces are used to ensure that the equipment is reasonably accurate. Hardness test operators should be trained to carry out frequent 'test block' calibrations, with results recorded and signed.

Most hardness test equipment has the facility to vary the load, thus it is quite easy to report peculiar results because the load applied is different from that which the operator thought had been applied.

During failure investigation it is routine to conduct hardness tests on the micro- and macro-specimens, with other tests being carried out as required. The tests on the components could be carried out 'as received' and after surface grinding, close to and remote from the failed area. The results, with comments, would be submitted to the investigator.

These could show any of the following:

- Homogeneous results all within specification.
- Homogeneous results all outside specification: these might be high, explaining a brittle type failure, or low, explaining ductile faiure.
- Variable results which could add to the information already available, or result in further investigation.

All this information would be obtained, in conjunction with the visual and metallurgical information, at much lower cost than other mechanical tests or chemical analysis.

Hardness specifications

Material specifications commonly include hardness as one of the control parameters giving a range of figures. Occasionally a specification will give a minimum figure as the only requirement.

The specification should state the type of hardness test and the load to be used. Abbreviations are almost invariably used on drawings, and commonly in material specifications. Samples of these, with translations, are:

Rc 55–60	Rockwell C range 55–60	= approx. 630–715 DPN
RB 55–60	Rockwell B range 55–60	= approx. 105–115 DPN
HBN 500–520 (also BHN-HB)	Brinell hardness	= approx. 550–570 DPN
DPN 630–715 (also VPN-HV)	Vickers (Diamond Pyramid Number)	= 630–715 DPN
Scleroscope 54–61	Scleroscope	= approx. 410–480 DPN

It will be seen that considerable confusion can arise regarding the actual hardness unless all parties concerned are careful to record the type of equipment used with the load and the figures specified.

In the UK the tendency is to use DPN (VPN HV) or HBN (BHN HB). These give comparable results at the lower hardness range, but as the hardness increases, the HBN curve tends to diverge.

Americans tend to use Rockwell hardness numbers, which can lead to confusion unless the method of testing is stated. Rockwell C is approximately 10% of the DPN (HBN) figure while other Rockwell methods A, B and D are quite different. The following conversion chart is in common use and illustrates the potential problem.

In addition to converting different hardness test results this supplies the conversion to ultimate tensile strength for steels.

Table 9.1 Hardness and tensile strength values approximate conversion table

Brinell				Vickers	Rockwell				
10 mm, 300 kg		*Approximate equivalent tensile strength for steel*		*Hardness number, HV, DPN or VPN*	*120° cone*			*$\frac{1}{16}$ in ball*	*Sclero-scope*
Dia. mm	*Hard-ness no.*	*tonf/lin²*	*hbar*		*C scale 150 kg*	*D scale 100 kg*	*A scale 60 kg*	*B scale 100 kg*	
2.00	945			1250	71	80	87	—	—
2.05	898			1150	70	79	87	—	—
2.10	856			1050	69	79	86	—	—
2.15	816			1000	68	78	86	—	—
2.20	781			975	67	78	85	—	106
2.25	745	163	251.7	950	66	77	85	—	100
2.30	712	155	239.4	910	65	76	84	—	95
2.35	683	150	231.7	850	64	75	84	—	91
2.40	653	143	220.9	790	62	73	83	—	87
2.45	627	137	211.6	750	61	72	82	—	84
2.50	601	132	203.9	715	59	71	81	—	81
2.55	578	127	196.1	671	57	69	80	—	78
2.60	555	122	188.4	633	56	68	79	—	75
2.65	534	117	180.7	599	54	67	78	—	72
2.70	514	112	173.0	572	52	65	77	—	70
2.75	495	108	166.8	547	50	64	76	—	67
2.80	477	105	162.2	523	49	63	75	—	65
2.85	461	101	156.0	501	48	62	75	—	63
2.90	444	98	151.4	479	47	61	74	—	61
2.95	429	95	146.7	459	45	60	73	—	59
3.00	415	92	142.1	441	44	59	73	—	57
3.05	401	88	135.9	424	42	58	72	—	55
3.10	388	85	131.3	409	41	57	71	—	54

3.15	375	82	126.6	395	40	56	71	—	52
3.20	363	80	123.6	382	39	55	70	—	51
3.25	352	77	118.9	369	37	53	69	—	49
3.30	341	75	115.8	358	36	52	68	—	48
3.35	331	73	112.7	344	34	51	67	—	46
3.40	321	71	109.7	332	33	50	67	—	45
3.45	311	68	105.0	321	32	50	67	—	44
3.50	302	66	101.9	310	31	49	66	—	43
3.55	293	64	98.8	299	30	49	66	—	42
3.60	285	63	97.3	290	29	48	65	—	41
3.65	277	61	94.2	282	27	46	64	—	40
3.70	269	59	91.1	274	26	45	64	—	39
3.75	262	58	89.6	267	25	45	63	—	38
3.80	255	56	86.5	260	24	44	63	—	37
3.85	248	55	84.9	253	23	43	62	—	36
3.90	241	53	81.9	246	22	42	62	—	35
3.95	235	51	78.8	240	21	41	61	100	34
4.00	229	50	77.2	234	20	41	61	99	33
4.05	223	49	75.7	228	19	40	60	98	32
4.10	217	48	74.1	222	18	—	60	97	31
4.15	212	46	71.0	217	17	—	59	96	31
4.20	207	45	69.5	212	16	—	58	95	30
4.25	201	44	68.0	206	15	—	57	94	30
4.30	197	43	66.4	202	13	—	57	93	29
4.35	192	42	64.9	197	12	—	56	92	28
4.40	187	41	63.3	192	10	—	56	91	28
4.45	183	40	61.8	188	9	—	55	90	28
4.50	179	39	60.2	184	8	—	55	89	27
4.55	174	38	58.7	179	7	—	54	88	27
4.60	170	38	58.7	175	6	—	54	87	26
4.65	167	38	58.7	172	4	—	53	86	26
4.70	163	37	57.1	168	3	—	52	84	25
4.75	159	36	55.6	164	2	—	51	83	24

Table 9.1 contd

Brinell				Vickers	Rockwell				
10 mm, 300 kg		Approximate equivalent tensile strength for steel		Hardness number, HV, DPN or VPN	120° cone			$\frac{1}{16}$ in ball	Sclero-scope
Dia. mm	Hard-ness no.	tonf/lin²	hbar		C scale 150 kg	D scale 100 kg	A scale 60 kg	B scale 100 kg	
4.80	156	36	55.6	161	1	—	51	82	24
4.85	152	35	54.1	157	—	—	50	81	23
4.90	149	34	52.5	154	—	—	50	80	23
4.95	146	33	51.0	151	—	—	49	79	22
5.00	143	33	51.0	148	—	—	49	78	22
5.05	140	32	49.4	145	—	—	48	76	21
5.10	137	31	47.9	142	—	—	47	75	21
5.15	134	31	47.9	139	—	—	47	74	21
5.20	131	30	46.3	136	—	—	46	73	20
5.25	128	30	46.3	133	—	—	45	72	20
5.30	126	29	44.8	131	—	—	45	71	20
5.35	123	28	43.2	128	—	—	44	69	—
5.40	121	28	43.2	126	—	—	44	68	—
5.45	118	27	41.7	123	—	—	43	67	—
5.50	116	27	41.7	121	—	—	43	65	—
5.55	114	26	40.2	119	—	—	42	64	—
5.60	112	25	38.6	117	—	—	41	63	—
5.65	109	25	38.6	115	—	—	40	62	—
5.70	107	24	37.1	113	—	—	40	60	—
5.75	105	24	37.1	111	—	—	39	59	—
5.80	103	23	35.5	109	—	—	39	57	—
5.85	101	23	35.5	107	—	—	38	55	—
5.90	99	22	34.0	104	—	—	38	54	—

5.95	97	22	34.0	102	—	—	37	53	—
6.00	95	21	32.4	100	—	—	35	51	—
—	94			98	—	—	—	50.5	—
—	93			96	—	—	—	50.0	—
—	92			94	—	—	—	49.0	—
—	91			92	—	—	—	48.0	—
—	90			90	—	—	—	47.5	—
—	88			88	—	—	—	46.0	—
—	86			86	—	—	—	44.0	—
—	84			84	—	—	—	42.0	—
—	82			82	—	—	—	40.0	—
—	80			80	—	—	—	37.5	—
—	78			78	—	—	—	35.0	—
—	76			76	—	—	—	32.5	—
—	74			74	—	—	—	30.0	—
—	72			72	—	—	—	27.5	—
—	70			70	—	—	—	24.5	—
—	68			68	—	—	—	21.5	—
—	66			66	—	—	—	18.5	—
—	64			64	—	—	—	15.5	—
—	62			62	—	—	—	12.5	—
—	60			60	—	—	—		—
—	58			58	—	—	—		—
—	56			56	—	—	—		—
—	54			54	—	—	—		—
—	50			50	—	—	—		—
—	49			49	—	—	—		—
—	48			48	—	—	—		—
—	47			47	—	—	—		—
—	46			46	—	—	—		—
—	45			45	—	—	—		—

Note: Conversion of one hardness test figure to any other can only be an approximation. Conversion of hardness test figures to ultimate tensile strength test figures is also an approximation and will also vary depending on the metal being tested. No hardness test, therefore, can be used to accurately define the strength of a metal.

It must always be appreciated that hardness testing supplies extremely useful information but cannot be compared to micrometer measurements. With a calibrated micrometer numerous operators will invariably obtain exactly the same results, with any deviation investigated and remedied. Hardness test results will vary with the same operator on the same specimen, and between different operators and machines. This variation should not be excessive: a 5–10% spread is acceptable.

If a specification calls for a hardness of 248–286 DPN (241–281 HBN) then readings as low as 240 or as high as 295 DPN should not result in rejection. However, the investigator should be aware that the material is suspect if it proves to be close to the limit, and this could add to other factors identified during the investigation. The report should also make it clear that while the material might be outside specification, it would be most unlikely that any marginal difference would be the major contributor to the failure.

Calibration of hardness test

The good investigative laboratory will have at least one hardness test machine, probably several of different types. It is good practice to calibrate the hardness test machine before every test. The calibration will use a test block with a known hardness range and the results will be recorded along with any deviation from the certified hardness. Any readings significantly different from the calibrated test block results should warrant a further test block tested either above or below the hardness of the specimen. That is, it should be determined if the machine is reading accurately in the region of the load used at hardnesses lower and higher than the hardness specified.

The test blocks themselves should be re-calibrated at least twice per year either by an outside authority, or by checking with other machines using different loads.

INTERPRETATION OF RESULTS

Hardness tests give a measure of the ultimate tensile strength of the surface of the material and the abrasion resistance. The tensile strength will be reasonably accurate for steels, with less accuracy for other metals, and even less for non-metals.

Any material which is harder than another will scratch the surface of the other. A soft material cannot scratch a harder material no matter what pressure is applied, while materials of similar hardness can scratch each other. This applies equally to metals and non-metals.

Where hardness testing is used to identify the strength of a material, care must be taken to ensure that a homogeneous sample is used. Under these conditions the hardness result will give a reasonably accurate measure of the ultimate tensile strength of a metal. With steel there are charts which quite accurately translate the hardness test into tensile strength. It is thus possible using hardness tests to carry out a survey across a specimen where failure has occurred; for example, a macro-sample with a weld with variation in hardness will indicate if there are areas which have lower tensile than others and, therefore, which could be expected to fail at a lower tensile load. Likewise high hardness figures in a weld heat-affected zone (HAZ) would account for a brittle failure in this area.

Hardness testing will also give some indication, although not necessarily accurate, of the impact properties. Almost invariably the harder a material, the less resistant it is to impact.

There is a greater variation when interpreting non-ferrous hardness test results, but it remains a fact that higher hardness represents higher tensile strength, and certainly better abrasion resistance.

Non-metallic materials in general obey quite different rules. Thus, for example, a soft rubber or plastic/polymer would be shown to be infinitely hard according to a Brinell test because of the lack of an indentation in the test. Specialist techniques are available which give reasonably accurate, reproducible results.

By the time hardness test results are available, the investigator will have completed the initial diagnosis, visual examination and macro- and micro-examinations, which will have given an indication of either a homogeneous structure, or various other possibilities.

Once the hardness test results have been obtained it should be possible to decide whether there is sufficient information to produce a report, or to carry out further work.

The investigator must always keep in mind the reasons why failure investigation has been requested and this should of course be part of his or her remit. If the investigation up to the stage of hardness testing has clearly shown that a material defect exists, then it should be the client who decides whether or not further work such as a mechanical test, chemical analysis, etc. is required. The information already available should explain why the problem occurred and if the client is the user of the equipment, it should be queried whether or not further expense is justified in carrying on to find out where the producer or user of the equipment went wrong.

To summarize, hardness tests as part of an investigation will identify firstly whether a metallic material is within the range of tensile properties which would be expected, secondly whether or not the material is homogeneous regarding hardness, and thirdly whether local areas of either too low or too high hardness have been the cause of, or contributed to, failure.

TYPES OF HARDNESS TEST

Following is a list of different types of hardness test, with information on the basic technique, and possible problems with the equipment, and reasons for using one type in preference to another. There are six methods of hardness testing.

Brinell Type (HBN, BHN, HB)

This method uses a hardened ball with a known load applied and impresses itself into the surface being tested. This will result in an indentation where the larger the indentation, the softer the material. The ball diameter and load applied can be varied.

The diameter of the ball impression is measured and this is used with a reference table to give the Brinell hardness number. (It is good practice to record the hardness figure obtained followed by the ball diameter and load applied in case of queries.)

The problem with the Brinell technique is firstly that it is relatively slow in application, requiring the forming of the impression, finding the impression, and then measuring this with a mirometer-type eyepiece. Secondly, as the hardness of the piece being tested increases, the ball will tend to distort. It should be obvious that once the hardness of the material being tested equals or is greater than the hardness of the ball, then little or no impression will be made and therefore it will appear to be an infinitely hard material.

The Brinell test is commonly applied in production and there are some types which can be used as portable hardness test machines but it is not commonly used in investigative laboratories. The 'Poldi' is a portable Brinell-type machine which is seldom used, having been replaced by the Equo-tip machine.

Diamond Hardness Testing – Vickers Hardness Test (VPN, DPN, HV)

This is commonly known as the Vickers hardness test and uses a diamond which for all practical purposes is an infinitely hard material and will cut or mark the surface being tested.

A predetermined load is applied to the indenter resulting in a diamond impression. The distance between the points of the diamond is measured and a chart used to convert this ocular reading into a 'hardness number'. The difference in length between the two diagonals of the diamond impression should not exceed 10%. This is now the test most commonly used in laboratories and is the technique to which other machines are calibrated.

The load applied can be varied and this can be a problem in that unless care is taken to ensure that the chart for the correct load is used, nonsensical readings will be obtained. This is a good reason for the habit of recording readings and loads; also ensuring routine checking of the machine with a calibrated test block is recommended.

This is the standard laboratory-type hardness tester, and is used for surveys, spot checks and other requirements during an investigation. It is non-portable, and has some limitations on the size and weight of specimens. The Equo-tip machine (which operates on a different principle) is also commonly used but it is recommended that there is continuous monitoring by the Vickers test or some other method.)

Durometer

This measures the resistance to load, rather than the result of the load applied. This can therefore be used for rubber and plastic, where quite often when the load is removed there is no indentation and thus the material would appear to be infinitely hard.

Rockwell Hardness Testing (Rc, HRC, Rb, HRB, etc.)

This technique uses an indentor which can be a hardened steel ball or a diamond; the load used can be varied. The machine measures the depth of the impression produced. Thus the shallower the depth the harder the material, the deeper the depth the softer the material.

The Rockwell machine can give very rapid results and is probably the most commonly used machine on production lines, but because of certain problems it is seldom used for investigative type work in the UK. Rockwell hardness test figures, however, will often be those specified, and thus appearing on technical documents.

Scleroscope and Equo-tip

The Scleroscope hardness technique uses the fact that a hardened steel ball will bounce higher from a hard surface than a soft surface. This is seldom used either in production or for investigative type work.

The Equo-tip hardness tester uses the same principle with an indentor which is given a known energy to hit the surface and then rebound. The speed of the rebound through a coil is measured and this is shown as a digital read-out. The faster the rebound, the higher the figure. There are tables to convert the digital read-out to Brinell, Vickers and Rockwell hardness figures.

The machine is extremely portable and easily used but requires considerable care as any vibration either because of the thickness of the section or from extraneous sources can produce less than sensible results.

This is seldom the only machine used for investigative type work. It is an extremely useful unit to identify whether components have a hardness problem, which can then be quantified using, say, the Vickers machine. It is also useful for giving scans to identify hardness variation across a workpiece, and is the most common hardness tester used for site investigative work.

Moh – Scratch Hardness Test

This technique was developed by geologists who often have a series of rocks or other items with variable hardness to quantify and the Moh scale was devised for them, with diamond being at the top end of the scale and talc at the bottom.

The technique has been used for many years now in heat treatment where a hardened file is used to abrade the surface of a carburized gear or other component. If the file bites into the surface then the surface is softer than the file and has not been correctly hardened. If the file slips off the surface without cutting or 'catching' the surface, then the surface is harder than the file. This is used in production for very rapidly and economically identifying surface hardness.

It can be used in investigative type work where various surfaces are suspect. Hardened needles can be calibrated and then used to scratch the surface. It is obviously very portable, and can be used on surfaces close to a change in section where no other hardness test could reach.

This technique requires considerable experience as it is a matter of feel whether or not the needle-point is scratching or slipping. It can, however, supply useful information in the correct hands. There is a Moh scale available, but when scratch needles are used they are generally calibrated to the VPN range.

Micro Hardness Test

This is the Diamond Hardness Test using a standard metallurgical microscope, and an indentor using pneumatic or hydraulic pressure to apply the load. The impression is minute. The results are notoriously inaccurate with a wide spread. Provided the equipment is in good condition and correctly used, the technique can be a useful tool to produce a number of impressions over the chosen surfaces. These can then be scanned to identify any gross difference in the size of the diamond, and the situation examined further.

To summarize, hardness testing will commonly be involved in failure investigation. It can indicate problems where variable, excessively high or low hardness is identified, confirming the reason for brittle or ductile tensile failure.

Sensibly used it can find unacceptable hardness (tensile) variations from surface effects (case hardening, de-carburization, etc.), weld problems, work hardening and faulty heat treatment.

The results might require confirmatory tensile or impact tests but in many cases can satisfy the client.

Mechanical testing

10

Mechanical testing is where a controlled stress is applied to assess material strength. Some form of mechanical test will have been specified as a matter of course with high quality components or where any problem would result in injury or high cost resulting from failure with the design load applied.

TYPES OF MECHANICAL TEST

The test or tests will use specimens of specific size and shape, which will be tested for certain parameters. The term 'mechanical test' covers any or all of the following. Those in italics are tests which could be requested as part of a failure investigation. The others are design features which would seldom if ever be carried out at a failure investigation.

- *Bend test*, e.g. ductility, weld tests, 'spring back'
- Compression
- Crack tip opening detection (CTOD/COD)
- Creep test
- Drop test
- *Ductility* – see elongation, reduction in area, under tensile test
- Exemplar test – see rig test
- Fatigue test
- *Impact test*
- Proof test
- *Reduction in area* – see under tensile test
- Rig test
- Shear test
- *Tensile test* – proof, yield, elongation, reduction in area
- Torsion test
- Torque test
- *Yield* – see under tensile test

The failure investigator must appreciate the relevance of each test, and should only require a test to be carried out if it will aid the investigation.

Most of the tests listed would equally apply to metallic and non-metallic materials; some tests however will have no relevance to specific materials.

An investigator will use visual examination, hardness testing and micro-examination to be certain that the component lacks ductility. Whether or not this needs to be confirmed by tensile testing and impact testing should then be discussed. If an investigation clearly shows a ductile fracture, and the micro-examination with hardness tests proves that the wrong material has been chosen, it should not be necessary to carry out a tensile or impact test.

If a failed component has no mechanical properties specified, then the results of a test, while supplying information to an investigator, could ultimately be irrelevant.

There will, however, be occasions when a specification is available and a component has failed, which will make it necessary to carry out mechanical tests to identify whether or not the component is within specification to satisfy a client. The investigator, however, must take care that the reason for carrying out any test is related fully to the investigation as it is quite common that there is no relationship between low mechanical properties and the reason for failure.

Each of the tests is described. The reason why a test should be carried out as part of a failure investigation will be identified as will why any test could be irrelevant.

Bend test

This relatively simple test is normally carried out on production components similar to the part which has failed. The failure investigator will seldom be in a position to reproduce these tests in the component which has already failed. The failure analysis should have looked for any indications that the production test has been incorrectly carried out or interpreted.

The various bend tests and types of samples that are used are as follows.

Ductility test	–	bar, sheet, tube
Face bend test	–	weld test piece
Root bend test	–	weld test piece
Side bend test	–	weld test piece
Nick bend test	–	fillet weld test piece
Spring back test	–	yield (proof) test

Bend tests are generally used as a simple, economic method of proving some form of ductility or tensile property. A brief description of each with its relevance to a failure investigation is provided.

Ductility test The test specimen can be 'as received' or machined as a round, flat or square bar. Where flat or square bar tests are required, the radius along the edges must be controlled. The test pieces should represent as far as possible the component in service. Tube will generally be as received.

The test must specify the radius of the former to be used, and the angle of bend acceptable. It is also necessary to clearly indicate and define failure end point. Tube bend tests should further define the degree of flattening in relation to the angle of bend.

The failure investigator will only be interested in a bend test if the visual examination has indicated poor ductility. The investigator may then decide to prepare a bend test from the component. This should be tested without machining the surface, which will put the failed surface into tensile during bending. This simple test could show brittle material, or the presence of surface stress raisers as contributing to this failure. This might obviate the need for tensile and impact tests.

Weld tests Face, nick bend, root and side bend tests, usually as standard quality control tests, are made on pieces produced with the same techniques used for production welds.

It is essential that all edges have a controlled radius, which should be representative of the production component. The former to be used will be detailed and the angle of bend specified.

With the **face bend test**, the upper surface or cap of the weld is placed in tension. This is a critical test to assess 'undercut' problems which can act as a stress raiser either to initiate fatigue, or brittle fracture of heat-affected zone.

The investigation may find that failure resulted from a stress raiser resulting from undercut. The report could then suggest that bend testing should be carried out on new production components, and if the failed piece was of a critical nature, that similar units in service should be inspected.

A typical test piece is shown in Figure 10.1(a). The test piece is supported over a known length and a load applied to the root. This will identify any problem at the weld cap, such as undercut, or serious hardness in the heat-affected zone. It will also show serious weld porosity.

A **root bend test** is the reverse of the face bend as shown in Figure 10.1(b). The load is applied to the weld cap, thus applying a tensile load to the root. This will identify any lack of penetration or excess hardness in the heat-affected zone.

Side bend tests bend the specimen sideways and will show cracking in the heat-affected zone, if the hardness is excessive. The failure investigation will identify the cracks by the macro-specimen hardness test. Failure could be expected to be a brittle fracture, or fatigue propagating from the brittle zone at the tensile surfaces.

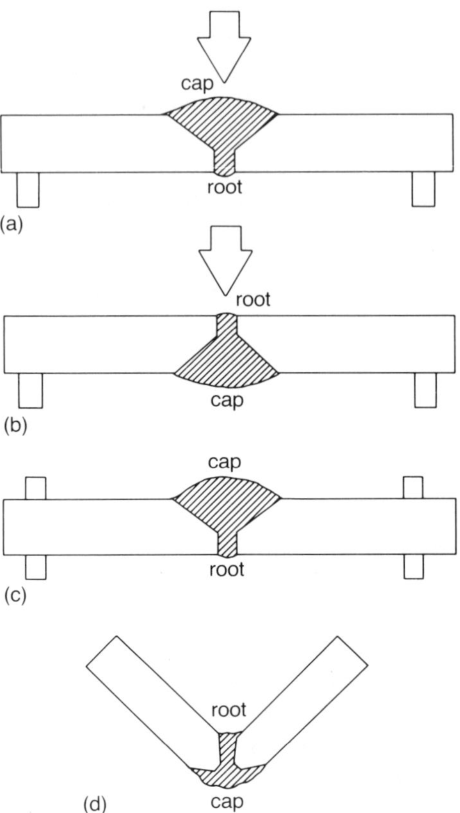

Figure 10.1 Weld test pieces for bend testing.

There could be occasions when a weld is associated with a failure that the investigator might carry out side bends. This could be to evaluate the results of hardness tests when there was any doubt if a weld failure was caused by weld problems, or a design fault.

The load is applied vertically to the prepared surface, that is to the top surface of Figure 10.1(c). A side bend test piece for a 90° angle joint is shown in Figure 10.1(d).

The **nick break test** is a production test used for fillet welds. The centre of the weld is saw cut to approx. 50% of the throat depth. The test piece is then bent to close the saw cut as shown in Figure 10.2. To be acceptable either there should be no failure, or fracture should only occur from the root of the saw cut. Fracture at either of the fusion areas is not acceptable.

The investigation will have identified at the visual examination or macro-examination whether or not this is a factor in the failure, thus re-testing would not be normal.

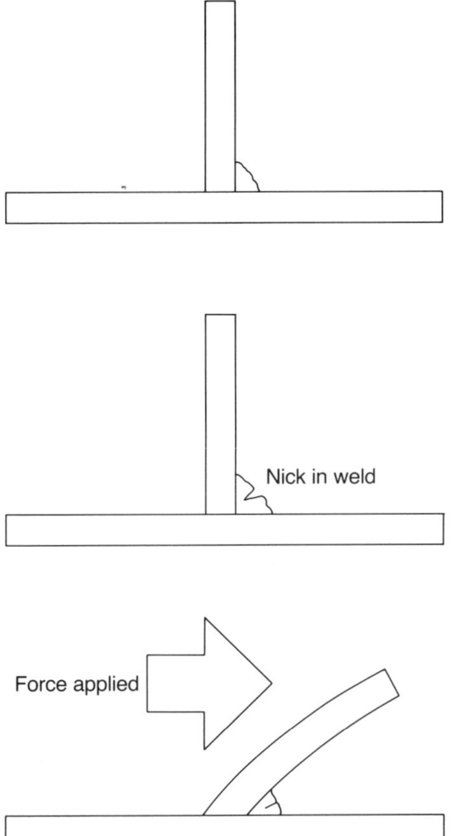

Figure 10.2 Weld testing using the nick bend test method.

Spring back test This is a simple test to indicate if the yield or proof strength of the material is correct. The test piece, either a round or a flat bar, or a piece cut from sheet is held in a vice and a load applied to bend the specimen.

The specification will detail the angle of bend to be applied, and whether or not it is held for a specified time. The load is then removed, and to be acceptable full recovery must be achieved.

The failure investigation will have identified whether or not the component had bent in service. The investigator would probably require to have a tensile test carried out if the hardness figures were reasonable. It is possible to have a material within specification for ultimate tensile strength but low in yield or proof strength.

There could be occasions when an investigator would carry out a simple spring back test to indicate rapidly and cheaply if the material was very weak.

Compression test

This is seldom specified, but when required there must be a detailed definition of what constitutes a failure. The yield or proof strength in compression for most metals is identical to that for tensile stress. When the tensile test piece reaches the yield point it will reduce in cross-section, thus further loading will continue to reduce the cross-section until a clearly defined fracture occurs.

With compression, when the yield point is reached, the specimen will increase in cross-section, thus a higher load is required to cause further distortion. Some specifications define failure as a certain permanent increase in cross-section area or diameter, others the first indication of surface bursts, or will accept that no bursting exists below a stated load.

Some metals such as cast iron, and other hard brittle materials, do not distort as the compression load is applied, thus it can be difficult, if not impossible, to identify the yield/proof strength in compression. Such materials will not normally show any evidence of compression failure by surface cracking or permanent distortion. Failure will be by collapse, which may be dramatic, if not dangerous, with brittle components such as ball-bearings.

Compression testing of concrete, brick and to a lesser extent stone, is however, a standard control test. Either a test cube, or a core sample, is tested to failure when the sample collapses.

Compression is a rare mechanism of failure with metals and will have been identified at the visual examination. Hardness tests and micro examination will give the investigator a clear indication of whether or not the material complies with the expected tensile properties, and it would be most unlikely that any confirmatory compression test would be required. But with non-metals such as concrete, rubber or other polymers, compresssion is a more common failure mechanism and reason for performing compression tests.

Rubber or plastic/polymer components might also have test pieces produced to identify, for example, the compression load required for permanent deformation.

Crack tip opening detection (CTOD)/crack opening detection (COD)

Both terms apply to the same test which has been recently devised to supply information on the ability of a material, very often associated with welding, to resist crack propagation.

Basically the test piece has a notch of a specified size cut from one surface. The test piece is then subjected to cyclic tensile loading to propagate a fatigue crack from the nose of the saw cut. This is controlled to give a known distance from the nose of the fatigue crack to the other surface. The test piece is then bent with the nose of the crack in tensile stress. The degree of bend, and load to fracture, supplies information on the crack propagation properties of the metal being tested. This is obviously an expensive test requiring sophisticated technical equipment and expertise.

A failure investigation will have found from the visual examination whether or not the fracture was brittle. It may have been necessary to carry out impact and tensile tests to quantify the situation. There should rarely be any necessity to carry out COD (CTOD) tests either on the failed component or on made-up test pieces.

If the visual examination shows a brittle fracture, and the impacts and ductility are satisfactory, some other unknown is likely to exist, requiring further investigation. This could be a low temperature effect, or the result of impact with stress raisers present.

These COD (CTOD) tests are now more commonly applied at the design stage, rather than production quality control or failure investigation.

Creep test

Creep is the elongation or stretching of metals with load applied significantly lower than the yield or proof strength. Metals such as lead, and to a lesser extent tin, creep at room temperature under their own weight. This takes considerable time.

Other metals require an increase in temperature before creep is identified. Creep is therefore temperature/time dependent – the lower the temperature the longer the time, the higher the temperature the shorter the time, before creep is identified.

Creep is found by a permanent elongation. Once a load is applied to the metal at a high temperature, creep will be accelerated. Creep testing is different from tensile testing. With hot tensile testing the test piece is taken to a specific temperature, tested at that temperature and the reduction in tensile properties are identified. Creep testing requires the materials to be loaded at a temperature and held at that load and temperature for periods up to years. Until the quite recent past, it was not possible to predict creep properties, but there is now sufficient data to predict the load expected to result in creep for a specific temperature and time. Thus creep testing is seldom carried out except on new materials.

The investigator will almost always be aware of when creep is a potential mechanism. The investigation will have identified certain characteristics at visual examination, and the investigator will then have carried out a micro-examination with specialized etching technique to identify the presence of creep voids, which will prove that creep is a factor in the failure.

It must be appreciated that creep failure is the result of temperature, time and load. The investigator cannot give a positive indication of which of these three factors has contributed most to the failure, but in some failures there could be features pointing to excess temperature, or load, and the client should be aware of the time the component has been in use.

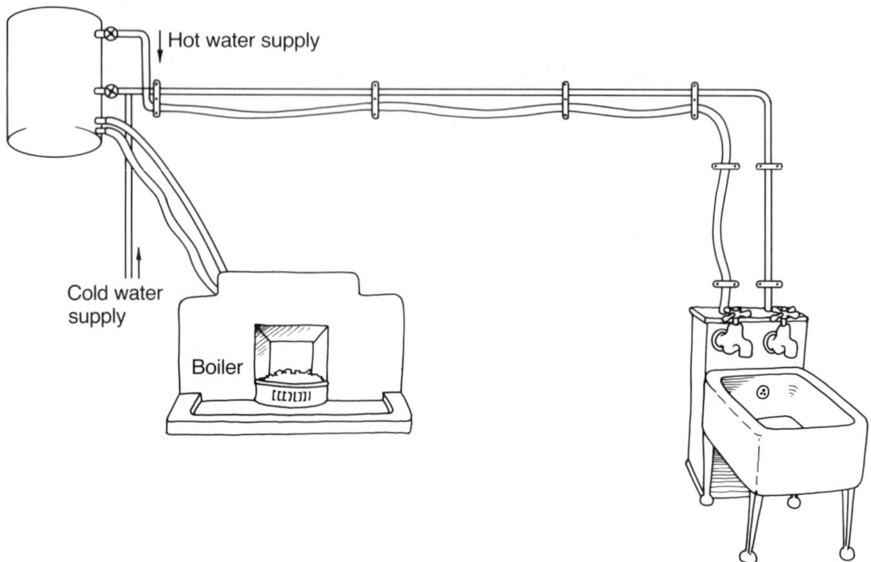

Figure 10.3 A familiar sight with domestic hot water systems prior to the use of copper piping: straight cold water pipes and sagging hot water pipes due to premature failure of the lead material through creep.

A familiar sight in the days of lead water pipes was the noticeable difference in creep of hot pipes compared with cold: indentical pipe runs would portray 'sagging and bowing' in the hot water pipes but not the cold water pipes, as illustrated in Figure 10.3.

Drop test

This is sometimes specified for electronic, hydraulic or pneumatic equipment to simulate mishandling. It is necessary, therefore, to specify exactly how each assembly will be assembled and packaged and exactly how it will be dropped, from what height and whether it will land on a corner or some other aspect. The type of landing must be specified and then the type of test which must be carried out after the drop. There will be a specified number of drops, sometimes from varying heights, with testing applied between the drops.

This test will only be carried out as part of failure investigation if there is indication that the failure is a result of impact or dropping, and to identify whether or not the component is capable of withstanding the standard test.

It might, however, be part of the recommendation of a failure report that drop testing should be made part of the specification, with information on this included in the discussion section of the report.

Fatigue test

This is the application of a controlled fluctuating or cyclic tensile load to a test piece, generally an assembly or an assembled component.

Information on the fatigue strength of all metals and alloys is available to the design engineer as either the endurance limit for steel, or the fatigue strength at a certain number of cycles for other metals. Thus, provided the component is loaded at less than the endurance limit or fatigue strength, failure will not take place unless the component is overloaded, or there is a stress raiser or the number of cycles has been exceeded.

S/N is the abbreviation for stress/number of cycles. What is known as the S/N curve is shown in Figure 10.4.

The design engineer has information regarding the fatigue strength of most metals. A rough rule of thumb is that this is approximately half the ultimate tensile strength. It will invariably be less than the yield/proof strength of the metal. The fatigue strength calculation to produce the S/N curve assumes that the load applied is evenly distributed across the stress surfaces.

The investigator who has found fatigue will look for stress raisers and consider the ratio of fatigue cracking to final failure. If there is an absence of stress raisers, and the fatigue propagation is significant, that is more than 50% of the cross-sectional area, then there could be discussion required regarding the design safety factor. If the fatigue propagation is significantly below 50% then there has either been an overload or the design safety factor is suspect. This will require discussion but not fatigue testing.

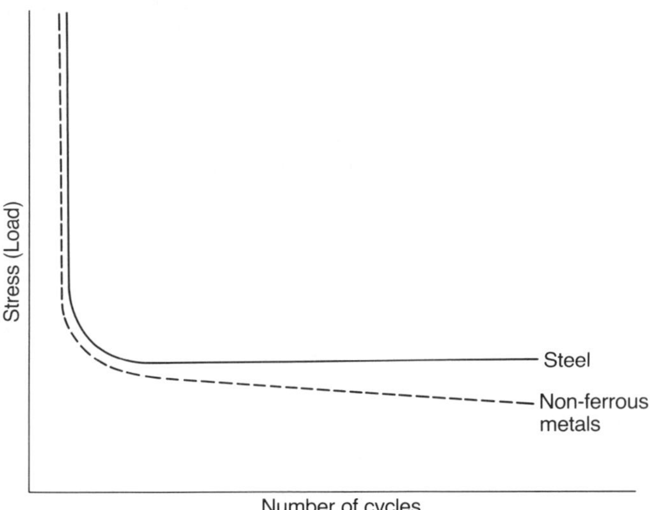

Figure 10.4 S/N endurance/fatigue curves for steel and non-ferrous metals showing stress vs. number of tensile load cycles.

Impact test

This requires carefully prepared test pieces with the standard method of test at present being the Charpy. In the past the Izod was the most popular method, but this has been largely superseded.

While the metallurgist may have some ideas regarding the impact properties of a metal from its metallurgical structure, it is not possible to predict with any degree of accuracy results of this test.

The test uses square or rectangular test pieces notched at the centre. It must be appreciated that the test represents the impact with a notch and it is the notch impact sensitivity which is being measured. The impact sensitivity might be quite different where there is a smooth surface.

It is also important to realize that the location of the test piece is very significant. It is common to take the test piece from one-third below the surface, in line with the tensile test piece. With the tensile test the results can be used to calculate the tensile strength of the bar or component by calculating the cross-sectional area. Impact invariably initiates at a surface and thus impact test pieces should be taken as close to the surface as possible, particularly if they are to represent the material in service, or the failed component.

There is a case history of a component which had satisfactory impact properties one-third below the surface, these being above 30 joules, whereas the impact at the surface was 5 joules. The metallurgical examination showed that this was the result of large grains at the surface from faulty heat treatment.

Both the Izod and Charpy test methods normally require three test pieces. The reason for this appears lost in antiquity. Normally one bend test and one tensile test are all that is required, but almost invariably there is a requirement for three impact test pieces. It is suggested that a possible reason is that in the days of the round Izod test, when the specimens had the notch machined by milling which required considerable skill to ensure a homogeneous notch, it was necessary to have more than one test piece because of differences in the depth of the notch. With the modern technique of broaching, Charpy test pieces give accurate reproducible results.

Some test houses now are prepared to use two test pieces with a third called for if it is not possible to identify the reason for any difference between the results from two test pieces. These laboratories do not average the results of impact testing and would invariably investigate any deviation of more than 20% from the central figure of the three. That is, results of 45, 40 and 80 J or 85, 80 and 40 J would have the non-standard result investigated. This might require metallurgical investigation if no physical difference could be found. This would apply in particular to welds or castings.

The Izod Test used is a round specimen of 11.45 mm (0.45″) diameter with a notch of 3.30 mm (0.13″) deep. A variation on the Izod test for castings uses a round bar with a larger diameter and which is not notched.

The Charpy test piece is a standard 10 × 10 × 55 mm long. This must have a good quality finish, and have a notch exactly at the centre. This notch must be carefully prepared, is extremely critical and must be tested at the exact centre. Where material thickness precludes 10 mm thickness, it is allowable to machine to 7.5, 5, or 2.5 mm. The test piece must always be 10 mm wide, 55 mm long, and conversion tables are available to allow for the non-standard thicknesses.

The test piece is located horizontally with the notch vertical and on the side away from the pendulum. The pendulum has a striking edge, which is swung through a known arc and generally must be calibrated at least once per year. Great care is necessary to accurately centre the specimens as the slightest off-centre will give high results. It will be obvious that the more brittle the material, then the further the pendulum will swing.

Before carrying out any test it is normal practice to ensure that with no specimen in position, the pointer (which is driven by the swinging pendulum) will register zero, that is an impact figure of zero.

Brittle specimens with readings below 10J, and perhaps as low as 3–5 J, are not unusual. Impact figures above 30 J at the test temperature are normally accepted as being reasonable. Figures below 20 J are suspect regarding ability to withstand impact. The highest figure obtained with good quality tough steel would be 300 J (220 ft lb).

A conversion table is given in Table 10.1.

It must be appreciated that there will almost always be some scatter of impact results; thus conversion factors and tables must be used with caution.

The results are reported as joules, foot pounds (ft lb) or kilogrammes/metre (kg/m^1). The conversion factors are:

ft lbs	×	1.356	=	J
J	×	0.737	=	ft lb
J	×	0.1	=	kg/m^1
kg/m^1	×	10	=	J
kg/m^1	×	723	=	ft lb

One important factor with impact is that, unlike tensile properties, as temperature is decreased there is not always a straight line relationship. Tensile strength or hardness measured as the temperature drops strength will increase, and ductility, measured as elongation, will decrease, exhibiting a linear relationship. With non-ferrous metals, including austenitic stainless steel, the same effect occurs with impact tests, that is with decrease in temperature there is a drop in impact. But with most steels this does not occur and at a certain point in the temperature curve there is a

Table 10.1 Impact data conversion table

ft lb	*joules*	*kg/m[1]*	*ft lb*	*joules*	*kg/m[1]*	*ft lb*	*joules*	*kg/m[1]*
1	1	.14	35	47	4.84	68	92	9.40
2	3	.28	36	49	4.98	69	94	9.54
3	4	.42	37	50	5.12	70	95	9.68
4	5	.55	38	52	5.25	71	96	9.82
5	7	.69	39	53	5.39	72	98	9.95
6	8	.83	40	54	5.53	73	99	10.09
7	9	.97	41	56	5.67	74	100	10.23
8	11	1.11	42	57	5.81	75	102	10.37
9	12	1.24	43	58	5.95	76	103	10.51
10	14	1.38	44	60	6.08	77	104	10.65
11	15	1.52	45	61	6.22	78	106	10.78
12	16	1.66	46	62	6.36	79	107	10.92
13	18	1.80	47	64	6.50	80	108	11.06
14	19	1.94	48	65	6.64	81	110	11.20
15	20	2.07	49	66	6.78	82	111	11.34
16	22	2.21	50	68	6.91	83	113	11.48
17	23	2.35	51	69	7.05	84	114	11.61
18	24	2.49	52	71	7.19	85	115	11.75
19	26	2.63	53	72	7.33	86	117	11.89
20	27	2.77	54	73	7.47	87	118	12.03
21	28	2.90	55	75	7.60	88	119	12.17
22	30	3.04	56	76	7.74	89	121	12.31
23	31	3.18	57	77	7.88	90	122	12.44
24	33	3.32	58	78	8.02	91	123	12.58
25	34	3.46	59	80	8.16	92	124	12.72
26	35	3.60	60	81	8.30	93	126	12.86
27	37	3.73	61	83	8.43	94	127	13.00
28	38	3.97	62	84	8.57	95	129	13.13
29	39	4.01	63	85	8.71	96	130	13.27
30	41	4.15	64	87	8.85	97	132	13.41
31	42	4.29	65	88	8.99	98	133	13.55
32	43	4.42	66	89	9.13	99	134	13.69
33	45	4.56	67	91	9.26	100	136	13.83
34	46	4.70						

dramatic reduction in impact values over a relatively small drop in temperature. This is known as the 'knee' in the impact curve and to date cannot be predicted. It is known to be dependent on heat treatment, impurities and other unknowns. The efffect is shown in Figure 10.5.

Specially formulated steels such as those with high nickel, 7–10% and upwards, have been produced and have no effect on the tensile properties but ensure that reasonable impact properties (at least 30 J) are obtained as low as −200°C.

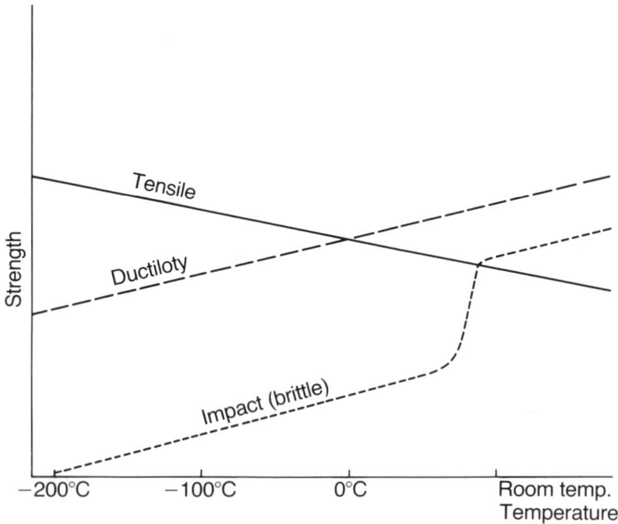

Figure 10.5 Variation of strength with temperature below ambient for steels showing the non-linear relationship of impact strength.

The failure investigator will have identified at visual examination whether a fracture is brittle or ductile. A ductile failure would seldom require impact tests being carried out, as impact has not contributed to the failure. Where, however, there is evidence of brittleness then it will be quite usual for impact tests to be specified. Care must be taken to ensure that the test pieces are machined to represent the metallurgical structure required and are as close to the fracture as possible.

If a material has impact values above 30 J, and certainly above 100 J, then it could be expected to resist crack propagation. If, however, there are large grains or micro-stress raisers at the surface, then these would be difficult to reproduce in an impact specimen. The nose of the test notch must of necessity be below the surface by at least the depth of the notch. These factors must be appreciated by the failure investigator. There can thus be surface effects with low impact values which are not identified with the standard piece. The metallurgist should be able to recognize these from the micro-structure, with evidence from the visual examination, and must interpret the findings on behalf of the client in the report.

Plastics/polymers can have the same type of test carried out. It can be expected that many plastics/polymers will have low or very low impact properties, while others will have very high impact resistance. The investigator should be aware that plastics/polymers can age in service, which can result in a dramatic reduction in impact strength from the properties ‘as produced’.

This problem does not arise with metals which do not age with time at ambient temperature.

Impact tests on wood are seldom required. One peculiarity reported with wood is that the force to fracture a longitudinal bend will be significantly less than that when a rapid impact is applied. It is suggested that with the impact the wood fibres absorb some energy, while with bending the fibres stretch.

Stone, concrete and brick cannot generally be expected to withstand impact. The sculptor, mason and bricklayer all use the brittle nature of these materials to produce, by impact, the shape or size required.

Proof test

This is different from the proof/yield strength. While of no relevance to the materials discussed in this book, it might be of interest that one of the earliest commodities subjected to quality control was whisky! The 'proof strength' was found by taking an equal measure of whisky, water and gun powder. If this could be ignited it was 'over proof'. By adding water until the mixture just failed to ignite, the degree of over proof could be found. Likewise by adding gunpowder to the original mixture until it could be ignited, the extent of under proof was found.

Normally proof testing is carried out on the components themselves under controlled conditions to find the proof/yield strength of the material. In general, but not always, the proof test load is calculated using the proof/yield strength of the material and the cross-sectional area. Certainly the proof test will never be used to apply a load which would take the component above its calculated yield strength.

One common component to be subjected to proof testing or proof loading is chain link and chains. Lifting chain in many cases is required to be proof tested during its service life. Anchor chain for ships is proof tested or proof loaded prior to going into service. In some instances ships' anchor chain is re-proof tested either after some specific action or after a period of time.

Proof testing or proof loading of other components and assemblies is frequently specified. This requires that the test itself is clearly defined and the load to be applied is specified.

Occasionally proof loading can be carried out where the component will in fact yield or stretch by a controlled amount. This ensures that the component is of the correct material and has been correctly heat treated to obtain the minimum mechanical strength specified as the yield or proof strength. As the proof load is applied, the component will stretch prior to the specified load being reached, resulting in some cold work which will increase the tensile strength.

There are variations of the proof test, for example high quality assemblies will require that the torque load applied to any bolts or studs will be such that this will take the component to above its yield strength. This means

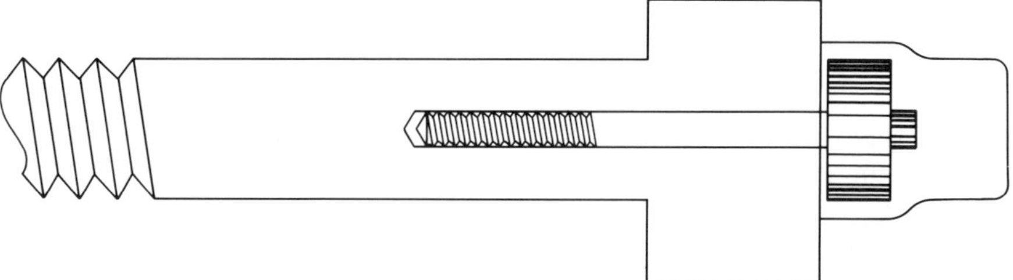

Figure 10.6 Rota-bolt method of proof testing.

that as the correct torque load is applied, the stud or bolt will continue to stretch and will clearly indicate that it is not of the correct strength.

Another variation of this is to fit a nut to a stud finger-tight, then to fit a second nut and using a fixture to apply a known load to stretch the stud to just below the yield strength. The amount of stretch is measured with feelers to find the gap between the first nut and the component. The first nut is then screwed to 'finger tightness' and the load removed. This means that there is a tensile load applied to the stud or bolt, which has stretched by a known amount. This eliminates the known problem of specifying a torque load which is designed to apply a known tensile load. The Rota-bolt is a development of this technique, shown in Figure 10.6.

Rota-bolts are proprietary components, supplied with a small hole drilled and tapped from the head down the shank. Into this is fitted a small 'bolt', with a known clearance and a cap between its underside and the upper surface of the Rota-bolt. This cap rotates freely. When the Rota-bolt is fitted and tightened it will stretch, thus reducing the gap. The correct tension has been applied when the cap does not spin.

Proof testing will never be required as part of a failure investigation, but if it is known that proof testing was required as part of the specification, and failure has been by tensile action, then a standard tensile test piece will clearly indicate whether or not the material was in the correct tensile/proof yield condition. If the material is below the specified yield or proof strength then it will be obvious that the proof test has not been carried out, or not been correctly carried out.

It is worth commenting that all anchor chain is proof tested before delivery, yet many failures have been investigated, all of which fractured with a load well below proof. Fatigue crack propogation can explain this, but many fractures in service are brittle, with no evidence of fatigue. This may be explained by the fact the proof load does not shock test the chain.

Rig Test (exemplar test)

Once a failure has been investigated and reported, it will often be possible to advise on how failure can be prevented. In most cases the recommendation will be either ignored or applied after discussion. Where the engineering in particular is complex, the suggested remedy may require to be tried out. This would also apply if a suggested material change was of a radical nature.

This trial would be rig tested to simulate operating conditions, which would probably be adjusted to accentuate or reduce the failure mechanism.

This is not in fact part of a failure investigation, but in most establishments the investigator would be consulted regarding the recommendation, who should be involved at all stages of the test to evaluate whether or not the suggested procedure has been correct, and to provide an examination of satisfactory parts.

Shear test

This test is sometimes specified for components which are required to fail in shear as a safety feature. It can be very difficult, if not impossible, to predict exactly the shear strength of a component unless a test has been carried out (see Figure 10.7).

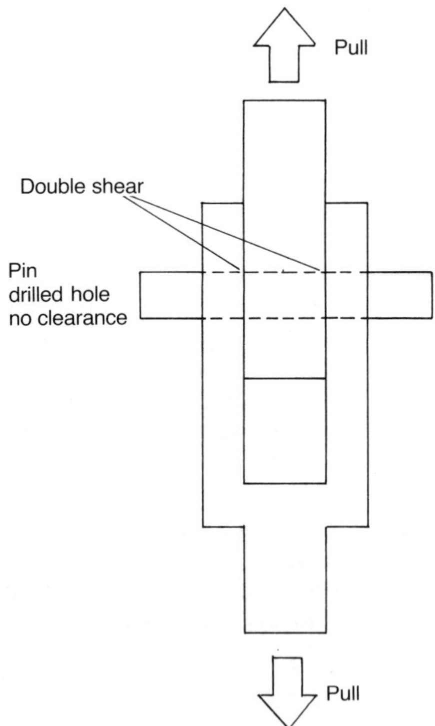

Figure 10.7 Method used for shear testing.

This test will normally be a double shear, that is, the component in the form of a pin will be assembled in a device. The clearance between the pin and the device must be very small, they should be a sliding fit, and the clearance between the moving parts of the fixture must also be very small to ensure that the component is placed in shear with no bending.

The assembly is fitted to a tensile machine and a load applied until failure occurs. On completion of the test it is essential that the component is examined visually to make certain that failure has been by pure shear with no evidence of any bending.

It would be unusual for this test to be carried out as part of a failure investigation, but there could be occasions when the client would require confirmation either that an overload had occurred or that failure was below that specified.

The shear strength of a material can be calculated from the tensile strength, but it is generally considered that because of machining tolerances, variations in heat treatments etc., for critical situations this test should be carried out on the actual component.

Tensile test

This is the most common mechanical test carried out by the failure investigator, and also during manufacture.

The production test piece should be made from the same cast as the component, and submitted to all the processes such as heat treatment, welding, etc. With high integrity components the test piece should be integral with the component, being removed prior to final inspection for testing. For standard components the test piece will be manufactured separately and may not quite represent the component because of a possible size difference at heat treatment.

One test piece is used to identify the yield or proof strength, the ultimate tensile strength, ductility as elongation and reduction in area. It would be rare for duplicate tensile tests to be carried out.

Test pieces are standard sizes, but where necessary, variations in size can be accepted. It is essential, however, that the gauge length, that is the length over which elongation will be measured, and the area in which failure may occur are parallel within very tight tolerances.

The original tensile test machine was a horizontal beam, pivoted at the centre, with the test piece attached between one end and the ground. Weights were placed at the other extremity until first the test piece yielded, and then fractured as further weights were applied.

The modern machine operates on the same principle with a mechanically driven beam applying the load to the test piece, which is fixed at the other end. The load applied and movement of the beam are recorded, thus producing the stress/strain curve. The yield strength (proof point) and ultimate tensile strength are calculated from the loads recorded. Using the cross-sectional area, the yield and tensile strength can be calculated.

With plain carbon steels and to a lesser extent alloy steels, there will be a well defined yield point. If the load-indicating pointer is carefully examined it will be seen that it will either stop moving or in some cases drop back as the load is continued to be applied. This is the point at which the end of the elastic limit is reached. Up to this point, if the load is removed from the test piece and the gauge length measured before and after, it will be found that the gauge length returns to its original size. In other words, the material acts in an elastic manner and full recovery to original size occurs when the load is removed. Above this point permanent stretch will have occurred.

With materials other than steel there is no well defined yield point and it is necessary to measure the extension of the test piece with the load applied and to plot these on a graph. Commonly it will be seen that above a certain load there is no longer a straight line but rather a curve. In practice it is difficult to identify the exact point at which deviation from linearity commences and therefore lines are drawn parallel to the elastic line. Calculations are then made regarding the load at 0.1%, 0.2% or 0.5% along the horizontal axis, strain (stretch). It is generally accepted that 0.5% proof strength is the yield strength.

Most tensile machines now plot graphically the load applied and the amount of stretch, that is the stress (load)/strain (stretch) curve is produced automatically as the test is carried out, as in Figure 10.8. The difference between 0.1 and 0.2% proof load is quite insignificant and probably in most instances within the limits of error for the test.

The results for the ultimate and yield/proof test should always be quoted in the units for which the tensile machine has been calibrated. That is, if the machine is calibrated in imperial tons, then the results should be reported in ton/in^2. The results can then be converted to the language to which the client is accustomed, e.g. 'tons force per square inch'. It might be a legal requirement that the original data is recorded and reported prior to conversion.

It is the yield/proof strength of a material which is the most important quantity within the subject of mechanical testing. Provided a material is used within its yield/proof strength then it will neither distort nor fracture.

The ultimate tensile strength (UTS) of any material is of no real significance either to the designer or to the failure investigator. It must, however, be appreciated that any measurement of hardness will be related to the ultimate tensile strength, not the yield/proof strength. It is common for hardness to be the tensile strength factor recorded on drawings.

The ratio of yield to ultimate tensile strength is significant as this can give an indication of the ductility which could be expected, measured in terms of elongation. With a low yield/proof strength and high ultimate tensile strength figures, it could be expected that there should be a reasonably good elongation figure, certainly above 20%. If, on the other hand, the yield/proof strength and UTS are close together then the elongation figure will be low, and this could account for a brittle failure.

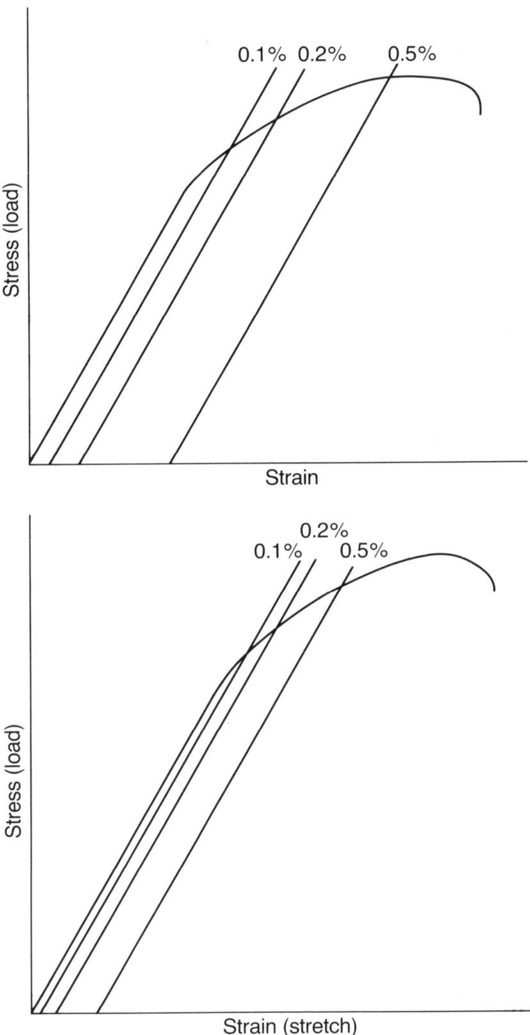

Figure 10.8 Typical stress/strain graphs produced during tensile testing. Percentages shown are differences in gauge length before and after fracture.

Elongation This is found by accurately measuring two points on the gauge length. These must be marked in such a manner that they can be readily seen before and after fracture, with care being taken that they do not act as a stress raiser resulting in the test piece failing at one of these points.

With round test pieces the gauge length will be related to the diameter of the test piece. There are two basic measurements for round specimens, one being 4 times the square root of the original cross-sectional area, the second being 5.65 times the square root of the cross-sectional area. These appear in specifications as $4\sqrt{So}$ or $5.65\sqrt{So}$.

The difference in calculated elongation for the same material can be significant, the translation being given in Table 10.2.

Table 10.2 Comparison between elongation values obtained on gauge lengths of $4\sqrt{So}$ and $5.65\sqrt{So}$

Elongation (%)	
$4\sqrt{So}$	$5.65\sqrt{So}$
8	5
10	7
12	8
14	10
15	11
16	12
17	12
18	13
20	15
22	17

It is of interest that the American system is less complicated, using the formula: 4 times the diameter (4D). This happens to be 2 inches for a 0.5 inch diameter tensile test specimen.

Where sheet material, or a square or rectangular bar is involved, the gauge length is marked by scores or light punch at regular intervals, for example: 1 inch or 25 mm intervals. The fractured test piece is brought together and the gauge length of either 2″ or 50 mm is remeasured and the elongation calculated as a percentage of the amount of stretch with respect to the original gauge length.

A variation of this is where ductility prior to 'necking' occurs. This can be a useful feature where sheet metal is to be subjected to any deep drawing operation. It requires a long test piece, with ten points marked 1″ (25 mm) apart. The test piece is broken, the two pieces brought together, and the increase in length over 10″ (250 mm) measured. The increase in length over the 2″

(50 mm) where fracture occurred has its elongation figure subtracted from the total length. The percentage elongation of the unfractured 8″ (200 mm) portion is then calculated.

It is not always appreciated that fracture can take place below the yield/proof strength or ultimate tensile strength, if the component, or part of the component, is taken above its elongation. Thus a bar which has a segregated metallurgical structure can have a low elongation at the centre – less than 5%, and satisfactory elongation at the surface – above 20%. If the bar is given a tensile load along its length, and the load results in stretching (elongation) or more than 5%, then fracture can be expected with failure initiating at the centre. The tensile load to cause failure would be below the yield strength of the material.

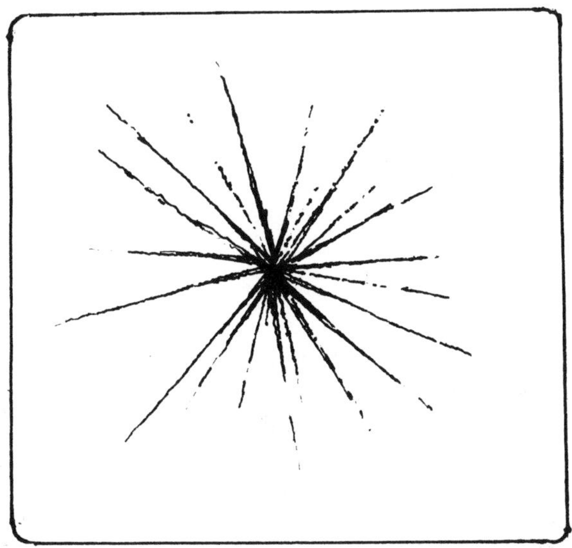

Figure 10.9 Unusual tensile failure initiating at the centre, not the surface. This was the result of carbon segregation, with the centre having a lower ductility than the surface.

Reduction in area (R in A) This is a measurement of ductility, much less significant than elongation. The diameter of the test piece is measured prior to testing. The broken test piece then has the two pieces brought together as for measurement of elongation, and the minimum diameter found with a pin micrometer. The percentage difference between these figures is reported as a percentage reduction in area, as shown in Figure 10.10.

To comment on this test, firstly there is the obvious difficulty of accurately measuring the minimum diameter on a fractured specimen. Secondly,

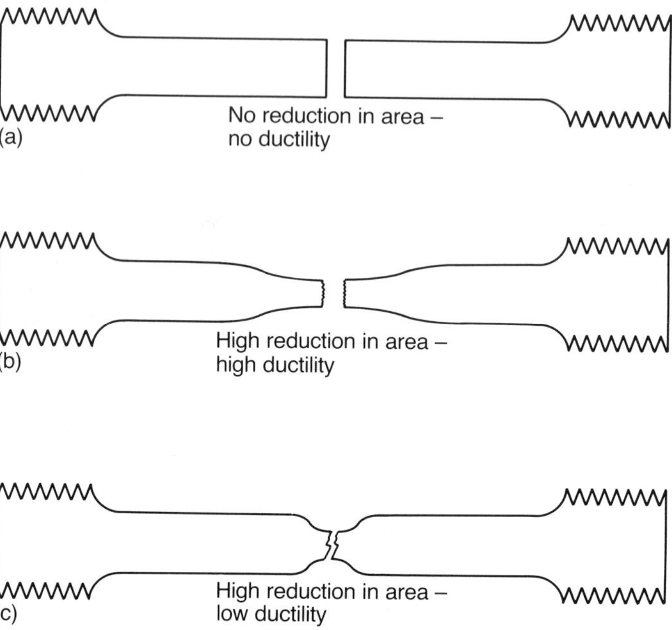

Figure 10.10 Assessment of reduction in area of testpiece.

to some extent, depending on the test conditions, the length over which the reduction takes place can vary. It will be appreciated that the longer the length over which the reduction in area exists, then the lower will be the percentage reduction in area.

There are some specifications, particularly with sheet metal components, where the reduction in area associated with the fracture is not included in the elongation calculations. The specimen is marked in equal increments along the length.

After fracture, the one or two increments where fracture occurred are rejected, and the increase in length of the others used to calculate the percentage elongation.

However, the failure investigator will have seen from the visual examination of the fracture whether or not necking has existed and will use this as useful information, the extent of necking indicating ductility at the fracture.

Torsion test

Torsion, which is a form of shear, is sometimes required to be evaluated. This will generally be the angle achieved under a known load. It generally requires a special fixture and discussion between the client and the mechanical tester. Use can be made of a standard tensile test machine, or application of dead loads.

Torque test

This test will seldom be specified for raw materials, but might be part of a specification for components. It would require that a detailed specification is produced indicating where the component is held, where the torque load is applied, and the degree of twist which will be accepted before distortion occurs. In most cases it will be quite difficult to carry out the test without special equipment to prevent any bending occurring while the torque is being applied.

A failure investigator identifying that torque is the mechanism of failure would require to carry out a hardness test initially and perhaps a tensile test to identify whether or not the material was within the specification limits.

If the hardness test identified that the material is below the tensile strength expected and the metallurgical examination indicated high ductility, low strength material, then no further work would be required. If, on the other hand, the material was found to be within specification in all respects then the client might wish more information on the degree of torque which would initiate failure. It is theoretically possible to identify the torque strength of a material from its proof strength and tensile strength. It would thus be unlikely that a torque test would be required.

There is a case history in the shear section where a torque overload was applied resulting in a pure torque failure. This led to a change in the control conditions of the equipment involved. No torque test was carried out.

Yield test

For information on yield strength see the earlier section on tensile test (proof strength).

11 Chemical analysis

Chemical analysis of metals, alloys or compounds is frequently included in failure reports. Most materials used, metallic or non-metallic, have some form of specification and this commonly includes chemical analysis, although this is less frequent with non-metals than with metals.

This chapter discusses the reason for chemical analysis being carried out and recommends that it should be performed only if it is relevant to the investigation. Chapter 13 on Materials information supplies more information on why any element is present, and there is some duplication of information. As stated in Chapter 3 on Report writing, it is essential that the investigator interprets the results of any analysis, and explains to the client the reason for the analysis being carried out.

Chemists, originally known as alchemists, have been a profession for very many years – back to the Middle Ages or beyond. (Metallurgy in the UK is considered a branch of chemistry). Most of their time was spent in looking for the 'Elixir of Life' or the magic potion which would turn metal into gold. While the chemist has not yet identified the Elixir of Life, the modern atomic chemist could, if so desired, turn some metals into gold. The cost of doing this far outweighs the value of gold but it can now be achieved.

ANALYTICAL CHEMISTRY

Until about 1960 chemical analysis of the type discussed in this book was carried out by what is known as 'wet chemistry'. Basically this meant that with a few exceptions the metal or alloy was dissolved. It was then either put through various reactions which resulted in a precipitate forming, being filtered off and being weighed to give the percentage of a specific element present or the dissolved metal was subjected to various chemicals which produced a colour change. This was then 'titrated' with another chemical which would react with the chemical in solution and when this was used up there would be a dramatic loss of colour or further colour change, creating a measurable end point from which element concentrations could be determined.

A limited number of elements, carbon in steel being the classic, were analysed by combustion methods. The finely divided metal, normally with a flux, was burned in a stream of oxygen to produce carbon dioxide. The volume of carbon dioxide was measured by dissolving it in caustic soda solution.

Wet chemistry required considerable skill and care. Where accurate results were required, estimations would be carried out in duplicate or triplicate. The chemist was assessed at frequent intervals by the use of standards where metals with a known make-up were submitted for analysis, and the results had to comply with those of the known alloy. Most chemists would accept that there were so many variables involved that a certain percentage of error must exist.

A major problem, which still exists with modern analytical analysis, is the sample supplied. If this is contaminated, the results will be meaningless. This can be a serious problem with investigation of samples, particularly where corrosion, contamination and abrasion are involved, or the samples are collected under less than perfect conditions. Where samples are taken to identify the make–up of metals, great care is required to ensure that the sample represents the relevant material.

Visual examination, with macro- and micro-specimens, may have identified differences which are relevant to the failure. Random samples for analysis under these circumstances could result in confusion. The investigator and chemist should discuss the situation and ensure that the sample or samples chosen represent accurately the failure and, if necessary, the basic material.

INSTRUMENTAL METHODS OF CHEMICAL ANALYSIS

At present almost all chemical analysis is carried out using instrumental techniques. It would be very difficult to find a laboratory now which uses as their basic method 'wet analysis' alone. There are still times when it is essential that wet analysis is used; for example chromium and iron are among the metals which can exist in two forms and only wet analysis can identify which form is present. However, the investigative metallurgist nowadays relies almost wholly on 'dry' instrumental analysis. This relies on two techniques.

Spark discharge method

This involves spark discharge on the surface, with the resultant spectrum analysed firstly for wavelength, which identifies the metallic element, and secondly for light intensity, which measures the amount of the element present.

The above comments regarding sampling are of particular relevance to this technique which can give nonsensical results if the surface for any of a number of reasons differs from the substrate.

There are a variety of spark discharge instruments which are confined to metallic analysis. The technique looks at only a small surface area, which usually has to be prepared. Other options using the same basic technique would be atomic absorption spectrometery (AAS), and induction coupled plasma (ICP). These take the sample into solution, which is then subjected to high temperatures, the resultant light-waves identified, and the intensity measured. Some of these methods can be used to analyse non-metals.

Electron beam method

The second technique now available uses the much higher energy of an electron beam. This technique can detect accurately all elements except hydrogen, helium, lithium, beryllium, boron and carbon with the standard machines, and even these can be detected in the more expensive, modern versions. This equipment is costly, but can carry out very rapid analysis.

The area analysed is very small and thus great care is required to ensure that the sample is representative of the problem being investigated. The electron microscope and its variations are briefly discussed in Chapter 1.

LIST OF ELEMENTS AND THEIR RELEVANCE TO FAILURE ANALYSIS

Table 11.1 contains a list of the elements commonly found in materials used in industry and commerce, which can be expected to be involved in failure examination.

Of the 103 known elements, some are radioactive, some short-lived and man-made, some quite rare or useless. The author considers the 38 elements listed in Table 11.1 to be those which will be encountered in industry.

It is appreciated that this is an opinion, with some elements included and others excluded, and those specific industries and individuals would offer a different short-list.

Those elements shown in bold in Table 11.1 feature in the 16 groups of compounds considered to make up the bulk of industrial materials. These are discussed further in Chapter 13.

There are four reasons why these elements should be present in an alloy.

Table 11.1 List of most common industrial chemical elements

1.	**Aluminium**	(Al)	13.	**Copper**	(Cu)	26.	Phosphorus	(P)
2.	Antimony	(Sb)	14.	Gold	(Au)	27.	Selenium	(Se)
3.	Arsenic	(As)	15.	Hydrogen	(H)	28.	**Silicon**	(Si)
4.	Beryllium	(Be)	16.	Indium	(In)	29.	Silver	(Ag)
5.	Bismuth	(Bi)	17.	**Iron**	(Fe)	30.	Sulphur	(S)
6.	Boron	(B)	18.	**Lead**	(Pb)	31.	Tantalum	(Ta)
7.	Cadmium	(Cd)	19.	**Magnesium**	(Mg)	32.	Tellurium	(Te)
8.	Calcium	(Ca)	20.	Manganese	(Mn)	33.	**Tin**	(Sn)
9.	**Carbon**	(C)	21.	Molybdenum	(Mo)	34.	**Titanium**	(Ti)
10.	Chlorine	(Cl)	22.	**Nickel**	(Ni)	35.	**Tungsten**	(W)
11.	Chromium	(Cr)	23.	Niobium/	(Nb)	36.	Vanadium	(V)
12.	**Cobalt**	(Co)		Columbium	(Cb)	37.	**Zinc**	(Zn)
	Columbium	(Cb)	24.	Nitrogen	(N)	38.	Zirconium	(Zr)
	see Niobium	(Nb)	25.	Oxygen	(O)			

1. The basic element making up the majority of the compound or alloy. It is suggested that there are thirteen of these, which are shown in bold in Table 12.1.
2. The elements are added to contribute to the properties of the material or alloy.
3. The elements are present in the ore or basic material and are not removed when the desired material is produced. These are known as 'residual' or 'tramp' elements.
4. The elements are added to reduce or overcome any problem caused either by the manufacturing process or by residual elements.

The same element could be found in an alloy or compound for any one or all four of the above reasons.

The remainder of this chapter takes each of the 38 elements which can be expected to be present for any one of these four reasons. The elements are listed in alphabetical order and an indication is given of why chemical analysis may be required as part of a failure investigation.

(1) Aluminium (Al)

This metal is one of the basic groups. There would never be any reason at failure investigation to call for confirmation of aluminium and its alloys. Aluminium will almost invariably be reported by a difference in specification. The metal is commonly used for cathodic protection of steel, and might require a qualitative check to differentiate it from zinc.

Aluminium is often present, however, in iron and steel, firstly as an additive for grain refinement purposes, secondly because it was present in the ore and cannot be fully eliminated, and thirdly as a de-oxidizing element. Failure investigation, however, would rarely require analysis for aluminium.

The metallurgist can estimate the extent and affect of non-metallics present, including aluminium oxide, either as residuals or as a result of de-oxidation. When used for grain refinement purposes, it is only soluble aluminium which is relevant and this is difficult to estimate chemically. A large grain structure could result in brittleness, though this could be for several reasons.

Aluminium is now used as a corrosion protection on steel. Thus qualitative analysis might be involved in a corrosion investigation.

Some duplex stainless steels have small amounts of aluminium as age hardening intermetallic materials. There may, therefore, be occasions when analysis would be requested.

With copper alloys there are a series of aluminium 'bronzes' where approximately 10% aluminium is present. Analysis for aluminium may occasionally be required by the investigator if, for example, there was a corrosion problem.

Aluminium is an inter-metallic element in many of the age hardening nickel chrome alloys. There could, therefore, be occasions where chemical analysis was requested by the investigator, but this would be to explain a specific mechanism of failure. The same comments would apply to the age hardening of titanium alloys.

Many of the zinc die cast alloys have significant quantities of aluminium, but it would be unusual for analysis to be required.

The only association aluminium might have with non-metallic materials would be as a coating. It may be necessary under some circumstances to identify whether or not a coating was aluminium. This would probably be qualitative rather than quantitative.

(2) Antimony (Sb)

Antimony is added as an alloying element to lead to increase the tensile strength and an investigator may therefore require analysis of antimony which would be cheaper than tensile testing, if a tensile failure was involved.

It would be rare that any other reason for analysis exists.

(3) Arsenic (As)

Arsenic is added to copper as a de-oxidizing agent and on very rare occasions an investigator might identify a need for chemical analysis on a copper specified as arsenical copper.

Arsenic is a well known toxic agent (as the oxide); thus forensic tests might be involved.

(4) Beryllium (Be)
A limited amount of beryllium is used in the nuclear industry, where the normal investigator would not be involved.

There are beryllium copper and nickel alloys where beryllium is an essential element and it might be necessary after a failure to require analysis. This should only be carried out in consultation with the client but can be cheaper than heat treatment trials and tensile tests.

There are some aluminium spray deposits with added beryllium; thus there could be a need for analysis depending on the mode of failure.

(5) Bismuth (Bi)
There are a series of low melting alloys, some of which have bismuth as the base material. They find use as fuses, and where for any reason low temperature casting is required. The investigator identifying a failure where the melting point was not within specification might require, after consultation with the client, to carry out analysis for the level of bismuth.

Bismuth can be used to improve machinability on austenitic stainless steels where analysis might be required, but this would be unusual.

(6) Boron (B)
Boron is added in minute quantities to certain grades of low carbon steel known as 'Fortiweld' to increase its hardenability during welding.

The investigator identifying a tensile problem would require to discuss with his client whether or not the relatively expensive cost of boron analysis is required. The competent metallurgist should be able to indicate from the failure mechanism, micro-examination and hardness surveys, whether the failure was boron associated.

Boron is a common fluxing agent in metal facing alloys. Analysis would be unlikely except with problems at the metal depositing operation. Some cast irons have boron added to reduce graphite formation; thus analysis might be required.

(7) Cadmium (Cd)
This is never used on its own or as an alloy but is commonly the corrosion protection of heavy duty steels, particularly in the aircraft industry, and to a lesser extent for marine use. It is not possible visually to identify whether or not zinc or cadmium has been used and thus chemical analysis would be required if a corrosion failure was found on a cadmium plated component. This would be qualitative, not quantitative.

There are some brazing alloys where cadmium is present to reduce the melting point or to increase shear strength.

It would be unusual to find these alloys at present as it has been identified that they are health hazards. It may, however, be that cadmium analysis is carried out on specific silver solders to guarantee that no cadmium is present. Qualitative analysis might be sufficient.

There is a copper alloy with 1% cadmium, where analysis might be necessary after a tensile failure.

(8) Calcium (Ca)

Calcium is used to control graphite formation in some cast irons. The metallurgist would identify whether or not the control had been successful. Thus analysis is not required at failure investigation.

Both aluminium and magnesium alloys can have calcium as a grain refining element. Calcium is a deoxidizer used when smelting beryllium, titanium, vanadium and zirconium. It is unlikely that a failure investigator would ever require analysis, and certainly only after discussion with the client.

(9) Carbon (C)

This is the essential element in almost all steels. With some, such as austenitic stainless steel and wrought iron, the amount should be as little as possible.

The investigator can obtain considerable metallurgical information from the amount of carbon present, and this is discussed in Chapter 13 on Materials information, and also in Chapter 8 on Micro-examination. There will, however, be occasions where chemical analysis for carbon is required. Even with modern methods there can be difficulty in obtaining accurate results and the investigator must be aware of this difficulty. This is often because of sampling difficulties.

This comment is particularly relevant at low levels, and thus where austenitic stainless steels have specifications of less than 0.03% carbon, alternative techniques to chemical analysis should be considered. One possibility would be to use the Strauss Test, which positively evaluates the possible consequence of above specification carbon. This is discussed in Chapter 13 on Materials information.

With many other metal alloys carbon may be present in small amounts which might necessitate analysis under certain conditions.

Carbon is also the major element in many non-metallic materials, but never under any circumstances would analysis be required.

(10) Chlorine (Cl)

This non-metallic element will never be present either on purpose or by accident in any metal or metal alloy. It is, however, an active constituent in corrosion and an investigator, to identify the source of corrosion, may require analysis for the presence of chlorine or chlorides. This may require extremely accurate analysis for low levels. This is discussed in the section on corrosion in Chapter 7 on Mechanisms of failure.

The investigator should be aware that handling components can result in chloride contamination, and that great care is required in preventing cross-contamination. There is thus the possibility that chlorides are identified and reported as contributing to the problem, when in fact they are the result of handling.

Many plastic/polymer materials contain significant quantities of chlorine, but it would be rare for analysis to be meaningful.

(11) Chromium (Cr)

This is a common and essential element in alloy steels either for hardenability, or for corrosion resistance and it is therefore common that analysis is carried out as a matter of routine.

The experienced investigator will, however, when identifying the mechanism of failure, know whether or not analysis would be relevant or useful. The micro-examination cannot, however, identify if the structure seen in the micro-specimen has chromium present.

Chromium when present as a tramp element can cause problems with mild steels and analysis may be required by the investigator having identified that there is a brittle structure which could be the result of high hardenability.

Chromium is used as a decorative, corrosion resistant or hard deposit on the surface of many alloys in addition to steel. It is not possible chemically to identify the difference between the different types of chromium deposits but qualitative analysis for the metal may be required if either corrosion or abrasion failure has occurred.

There are some chromium copper alloys used for electrodes where analysis might be required when the initial investigation has not found the cause of a tensile failure.

Chromium is present in nickel and cobalt alloys in significant quantities. It would be unusual for analysis to be requested except in special circumstances.

(12) Cobalt (Co)

This is one of the common metallic elements which appear in this book as the basis of a range of alloys. Some cobalt alloys are quite complex and there may be occasions where the investigator would decide it was necesary to identify the amount of cobalt present. Some alloys require a ratio of various elements to obtain specific properties.

Cobalt is used as an alloy in many steels. It has properties not unlike those of nickel but tends to enhance mechanical properties when present, particularly toughness. There could therefore be occasions when the investigator, in agreement with the client, would require analysis to confirm the reason for failure.

Traces of cobalt will almost invariably be found with nickel alloys. Cobalt is accepted as being harmless, in fact generally advantageous, in small quantities.

There are many magnetic materials which contain significant quantities of cobalt, the level of which could require analysis for a number of reasons.

Some of the more complex nickel chrome alloys contain significant quantities of cobalt as part of their specification. The failure investigator would require to identify the reason for the cobalt being present and then to assess whether or not the failure mechanism indicated that the cobalt level was too high or too low and how significant this was.

With tungsten carbide materials, cobalt is a common bonding element in the matrix. There could be occasions where the ratio of cobalt to the tungsten carbide could be relevant. It would be simpler and more economical to perform an analysis for cobalt than one for the more complex tungsten carbides, if the micro-specimens did not supply the required information.

Columbium (Cb)
See note under 'Niobium'.

(13) Copper (Cu)
Copper forms a range of alloys and is unique among the common metallic elements listed in that it is frequently the subject of analysis. The use of electrolysis enables the chemist to accurately estimate the percentage of copper, even above 99%.

Thus copper analysis might be required if, for example, the investigator requires to identify the difference between alpha and beta brasses. Until the advent of modern analytical techniques the majority of copper specifications gave the percentage of copper by estimation and of zinc by difference.

With modern techniques the chemist now has the choice of analysing for zinc or copper to identify the type of brass involved. In most cases analysis would only be carried out after discussion with the client.

The cupro-nickel alloys might require analysis for copper in preference to nickel under some circumstances.

Copper is added at approximately 0.5% to a range of steels known as Corten. These have enhanced corrosion resistance, the copper stabilizing the red/brown iron oxide. Thus the investigator with a corrosion problem on one of these steels would require analysis for copper.

With most steels, copper is an undesirable element as it tends to adversely affect ductility, but it will seldom require analysis.

With aluminium alloys, however, copper is one of the elements which form inter-metallic compounds, which are taken into solution and then precipitate in the 'age hardening' process. Thus when an investigator identifies tensile problems in aluminium, analysis of copper would be desirable.

It is possible to colour-anodize aluminium to give a surface appearance of copper or copper alloys such as brass or bronze. These can normally be identified by simpler means than analysis.

Zinc alloys sometimes have copper added to improve mechanical properties. Thus, like aluminium, if tensile problems have been identified at failure, analysis of copper might be desirable, but it is seldom carried out in these low cost materials.

There are a series of nickel alloys under the general name of Monel where copper is a significant element. It would, however, be unlikely that failure investigators would require analysis of copper.

With the nickel age-hardening alloys there are some alloys where the copper forms inter-metallics which take part in the precipitation hardening process. If the investigator requires analysis for copper, this should be carried out only after discussion with the client.

With non-metallic materials there will seldom be a need for copper analysis. It is clear that there could be very specific cases where copper either was suspected as a separate contaminant, or was part of a sophisticated product. The investigator would have this information. Copper is used as a surface coating on many electronic plastic/polymer components, but visual examination can identify if copper exists, and micro-examination would find its thickness.

(14) Gold (Au)

Chemical analysis for gold might be required when a contamination or corrosion problem was identified on an electronic component where gold was used for conductive purposes. The investigator would have identified a problem and might require to carry out a chemical analysis.

Gold is also used as a fuse material and it might be necessary to analyse for the percentage of gold. There are also brazing materials which use gold and again analysis might be required, after preliminary investigation.

At one time gold was used in the electronics industry, as a solder 'stop off', and at the soldering operation the gold was dissolved by the solder. At a certain concentration gold 'contamination' results in brittle solder joints. Analysis was required firstly to identify that the gold content was below the maximum acceptable, and secondly for assay purposes in order to find the value of gold in the solder. Gold is no longer used for this purpose.

Aluminium can be colour-anodised to give a very good appearance of gold. A simple weight test should be all that is necessary to differentiate gold from aluminium.

(15) Hydrogen (H)

This is the only gaseous metallic element at room temperature. Industrial materials in general, particularly metals, do not regard this as a desirable constituent.

There are various mechanisms of failure where hydrogen might be suspected of being responsible. Being an active gas, it disappears before it can be trapped for analysis. There may be rare occasions where a sophisticated failure has been identified and analysis might be required. This would make use of an electron microscope, the modern types of which can identify hydrogen.

Where hydrogen embrittlement is diagnosed or suspected, analysis for hydrogen would not be required.

With non-metals hydrogen is a common constituent, but it is very unlikely that this would ever be significant in a failure investigation.

(16) Indium (In)

This is similar to lead and finds limited use in industry. There are occasions where it will be specified either as a metal seal or as an electroplated deposit.

If an indium seal failed, it might be necessary to carry out qualitative analysis, as visual and metallurgical examinations could not identify the difference between lead and indium.

Where indium has been specifed as an electroplate deposit on bearings, if seizure occurs the investigator might require to identify whether or not indium has been present. Again this would be a qualitative analysis.

Indium is an alloy in low melting point alloys for fuses. If problems arise because of unscheduled failure (melting, or not melting when required) then analysis might be necessary. This would be quantitative.

(17) Iron (Fe)

This is by far the most common metal element used for industrial purposes. There are, however, no occasions where chemical analysis would be requested or required for any iron or steel product. Iron, where specified, will always be reported by difference, never by chemical analysis.

In some precipitation hardening aluminium alloys, iron will be present in small amounts, but it is unlikely that an investigator would require chemical analysis. Certainly it would require consultation with the client.

A maximum level of iron may be specified for other aluminium alloys but it would be unlikely that an investigator would require chemical analysis.

There is the possibility that corrosion contamination problems could be contributed to by the presence of iron. Iron would act as the cathode with aluminium being the anode, thus corroding. There could then be occasions where the investigator would require analysis of contamination or corrosive products to identify the presence of iron.

With the copper series, there are aluminium bronzes which contain an appreciable quantity of iron. Where the investigator identifies tensile-type failure, analysis of iron might be required along with the other alloy elements, but this would be unusual.

With nickel/chrome precipitation hardening and non-precipitation hardening alloys, iron is an essential element with some specifications and under some circumstances chemical analysis may be required.

Some precipitation hardening titanium alloys have iron as an essential ingredient in small amounts. It would probably be more economical to carry out analyses for iron and other essential elements if a tensile type failure had been identified as resulting from poor mechanical properties.

Chemical analysis in this instance would be more economical than sophisticated heat treatment trials. If the analysis showed all the critical elements to be within specification, it would be reasonable to assume that a poor tensile strength is the result of heat treatment.

(18) Lead (Pb)

This is one of the 16 metallic groups discussed in this book. There is no obvious circumstance where an investigator would require quantitative chemical analysis of lead or lead alloys.

There could be occasions where there would be a requirement to identify whether or not the material being investigated was lead or another soft grey metal. This could be carried out by qualitative analysis.

Lead is an additive of many steels, particularly low carbon mild steel and stainless steels to aid machinability. The metallurgical investigation would identify rapidly and economically whether or not there was a free machining constituent present, and also the form in which it was present. This would supply much more information to the investigator than whether or not the free machining constituent was lead.

Lead bronzes at one time were common bearing metals. While less common now, such alloys are still used and a bearing failure which could not be explained for other reasons might require analysis for lead content along with other elements.

There are a few aluminium alloys in which lead in small amounts is specified, but it would be unlikely that analysis would be required, as the presence of lead could be identified by metallurgical examination.

Most common solders contain lead in appreciable amounts of up to 40% or more. The ratio of lead to tin in critical soldering operations is important and thus an analysis of lead with or without tin would be required.

(19) Magnesium (Mg)

Magnesium is one of the common metals having a series of alloys listed in this book. The investigator of a magnesium alloy failure would rarely require quantitative analysis. If there was no information on the material involved, a qualitative check might be required to differentiate magnesium positively from other light white metals such as aluminium.

Some cathodic protective anodes are magnesium, and again a qualitative analysis might be advisable.

(20) Manganese (Mn)

Manganese is present in almost all steel specifications, as it is the most common deoxidizing element. It would be unusual for an experienced investigator to require chemical analysis. There are, however, high manganese steels – 6% up to 20% manganese – which have cold working properties. Failure could require chemical analysis, depending on the mechanism.

Manganese is an alloying element in aluminium and copper alloys. In most cases it is designed to increase the tensile strength, and in some cases to enhance the cold working capability. Thus manganese could be critical in supplying the reason for a tensile failure, and analysis could be requested.

There are nickel manganese alloys which have specific coefficient of expansion characteristics, thus analysis might be required.

(21) Molybdenum (Mo)

This is an expensive element but is commonly used in heat treatable alloy steels, and in some stainless steels. If there is mechanical failure with molybdenum specified then chemical analysis might be required.

Where stainless steel corrosion resistance is suspect then analysis for molybdenum would be required if type 316 had been specified.

There are now low friction coatings available based on molybdenum disulphide. Qualitative analysis might be required if abrasion or seizure has occurred.

(22) Nickel (Ni)

This is one of the common elements considered in Chapter 13. It is very seldom that nickel analysis will be carried out on any nickel alloy or on pure nickel.

Austenitic stainless steels have a minimum of 8% nickel, up to over 20% for heat oxidation resistant steels. If the steel is non-magnetic it would only be the higher nickel content steels which might require analysis if certain type of failure were found.

Duplex stainless steels invariably have a significant amount of nickel, which can be a critical element in corrosion resistance; thus analysis might be requested.

Where nickel is specified in heat treatable steels, and this applies particularly to carburizing steels, then its absence or being out of specification could be critical and chemical analysis would be required to identify whether or not the correct amount of nickel existed, particularly if a brittle tensile failure was identified, or if low core strength existed.

Nickel is found in some aluminium alloys but these are rather specialist and would seldom require chemical analysis.

In some copper alloys, in particular the cupro-nickels, the amount of nickel could be critical regarding the corrosion resistance and nickel analysis might be essential.

With aluminium bronzes the presence of nickel considerably improves the properties, and can only be identified by analysis.

With the cobalt alloys, nickel will almost always be present but this can be in variable amounts, and for various reasons. The need for analysis would depend on the failure mechanism.

23. Niobium (Nb)/Columbium (Cb)

There are no useful alloys of niobium. It is, however, used at low levels – less than 0.2%, in certain austenitic stainless steels. These are the weld-stabilized steels where niobium prevents the defect 'weld decay'. Chemical analysis would never be called for if this defect was suspected.

Using the Strauss test the tendency for weld decay to occur can be positively identified. This test uses an acidified copper sulphate solution in which the sensitised specimen is boiled for 72 hours. The specimen is then bent through 90°, and if no cracks appear there will be no weld decay. Sensitizing is achieved by heating the test pieces to 650°C for 0.5 hour.

Niobium is also used in low alloy weldable steels where it inhibits grain growth. There could be occasions where the investigator would require analysis for niobium, or to carry out tests for grain growth.

24. Nitrogen (N)

Nitrogen is the major constituent of air and thus any smelting operation is liable to have the end product contaminated with nitrogen. In general it is undesirable and the smelter takes care to eliminate the problem as much as possible.

If brittle tensile stress has been identified as the mechanism of failure, there must always be the possibility that nitrogen contamination in the form of brittle nitrides has contributed. It would, however, be very seldom that the investigator would specifically request nitrogen as an element for analysis.

Nitriding is used as a surface hardening/case hardening process. The metallurgist, however, can identify whether or not the case is present and is of the nitriding type and would seldom or never involve chemical analysis.

With some of the modern Duplex stainless steels, there is a specified nitrogen content which is essential firstly for the correct age hardening, and secondly to reduce pitting corrosion. An investigator faced with either low strength or excessive pitting on specific components where the same conditions did not affect other duplex stainless steels might require nitrogen analysis, which can be quite difficult without specialist equipment. If, however, the client required exact information on the reason for failure then chemical analysis would be more economical than sophisticated heat treatment or corrosion testing.

At temperatures above 250–350 °C titanium and its alloys have an affinity for nitrogen which could result in problems. Titanium has an equal affinity for oxygen, hydrogen and other elements in the atmosphere, and the investigator should not need to carry out analysis if the problem is temperature associated.

25. Oxygen (O)

Like nitrogen, oxygen is present in the atmosphere and is much more reactive; thus it presents more serious problems during any smelting or heating operation. It is the major contributor to atmospheric corrosion. The investigator examining contamination or corrosion as a problem would never want oxygen analysed but would be looking for the metallic element in the contamination or corrosion products.

All steels and wrought and cast iron will have oxides present as slag or oxide stringers and at no time will the investigator ever require to have chemical analysis for oxygen. While it is not possible to differentiate by metallurgical means between oxides, sulphides and phosphides, this is seldom relevant as any one of the three could contribute to a problem.

Aluminium has a great affinity for oxygen and thus aluminium and its alloys will have oxides present. The metallurgical investigator may have identified that oxides could have contributed or be the reason for failure. He or she would, however, never require to confirm this or to determine the quantity of oxide present.

With copper a very small amount of oxygen, measured as parts per million, can have a dramatic effect on the electrical conductivity.

If the investigator identified that conductivity or lack of it was the problem with copper, then an oxide might be responsible, but it would be very unusual that the amount had to be quantified. Certainly no chemical analysis for oxygen should be requested without confirmation of the client.

Nickel forms a stable surface oxide, and it would be unlikely that any investigation would require analysis at any time.

Other materials such as magnesium, titanium and zinc have an affinity for oxygen, but there is no history of requirement for chemical analysis.

With non-metallic materials, particularly plastics/polymers, and to a lesser extent wood and natural materials, oxygen at even warm temperatures can accelerate quite dramatically polymerization or ageing of the materials, but this would not result in a request for chemical analysis for oxygen.

There are, however, analytical techniques which can compare the chemical structure of organic compounds, and indicate the degree of polymerization which could be blamed on oxidation.

26. Phosphorus (P)

This is commonly listed in metal specifications, particularly for steel, at a maximum level. Phosphorus is present in the ores and remains in some of the metals.

It will be present in iron and steel as iron or manganese phosphide and will be seen by the metallurgist as a non-metallic constituent referred to as oxide or slag. The metallurgist can estimate firstly the total quantity of oxide or slag, and the form in which it is present, and thus chemical analysis is seldom or never required. It is a fact, however, that because it is part of the specification it is very often analysed and reported.

Cast iron commonly contains phosphorus as a phosphide and this can be recognized as a specific constituent, but there could be occasions when it has to be quantified by analysis.

There are alloys in the copper range known as phosphor bronzes which contain significant amounts of phosphorus. This is reputed to improve the bearing properties of the cast metal and thus there could be occasions where the investigator might require chemical analysis.

27. Selenium (Se)

This metallic element is occasionally specified to improve machinability and thus is found in free machining steels, particularly austenitic stainless steels. The metallurgical investigator can identify in the micro-specimen whether or not a free machining constituent is present in the approximate amounts and in the correct distribution. It would, therefore, be necessary for chemical analysis to be requested only if there were dispute regarding whether the constituent present was selenium or one of the other free machining constituents such as lead, sulphur or telurium. This wold be most unusual and thus the situation should be discussed with the client.

28. Silicon (Si)

This is a major constituent in non-metallic materials such as concrete, brick, glass and stone. Silicon on the Earth's surface in the form of silica

(SiO_2) is better known as sand or quartz. It would be most unusual to require analysis for silicon in any of these basic materials.

There are now, however, a series of materials based on this element which are analogous in many ways to organic materials. These are known as silicones and are discussed in Chapter 13 on Materials information.

In iron and steel any silicon remaining will be seen as oxide or slag and the experienced investigator will be able to assess whether or not this has contributed to any problem. Thus there should seldom be any need for silicon analysis to be requested except in exceptional circumstances. There are certain specialist steels and irons for which silicon is specified within limits, for example with the core material for transformers. The specialist investigator under some circumstances may therefore require silicon analysis. There are also cast irons with high silicon content for corrosion resistance for which analysis might be necessary.

Like phosphorus and sulphur, silicon is quoted in the majority of steel specifications but an investigator would seldom require an analysis. However, it is quite common for analysis to be carried out and reported as part of a routine.

29. Silver (Ag)

This will be most commonly found as an electroplated deposit. Where the investigator has identified a certain type of problem it may be necessary for analysis of silver which will generally be qualitative only.

There are a few other specialist products discussed in Chapter 13 on Materials information, notably bearings and fuses. The investigator will be aware from the mechanism identified and the specification or information regarding the purpose of the component whether or not chemical analysis would be required.

30. Sulphur (S)

Sulphur is found with many ores and thus finds its way into metals at the smelting operation. This applies in particular to iron and steel, and to a lesser extent copper and nickel.

Sulphur, with few exceptions, will be specified as a maximum. The skilled metallurgist can estimate whether or not the oxide/slag content of steel is within the limits which would be expected for low sulphur steels. Therefore, there should be no need for sulphur analysis, although as with phosphorus and silicon this is commonly carried out as a matter of routine.

Sulphur is also one of the constituents chosen when free machining steels are required. These would have the sulphur content specified as a minimum and a maximum, with the maximum being something like 5 to 8 times higher than for non-free machining steels. The manganese content of steel would be raised in proportion to the sulphur content.

Like the other free machining constituents, that is lead, selenium and tellurium, it is the particle size, distribution and quantity which decides whether the material is machinable or not. Therefore, analysis which will only tell the quantity of sulphur present should be carried out only in agreement with the client.

Sulphur as sulphur dioxide (SO_2), and to a lesser extent hydrogen sulphide (H_2S), is found in the environment and thus can contribute considerably to corrosion/contamination problems. There may, therefore, be occasions when the products of corrosion or contamination require analysis for sulphur to pinpoint the reason for the corrosion. Nickel and its alloys in particular are prone to sulphur attack, especially at higher than ambient temperatures.

Sulphur contamination can also cause problems with non-metals. This can be degradation of concrete and plastics/polymers. Analysis of the contamination would commonly look for sulphur.

31. Tantalum (Ta)

This has a few specialist applications where any failure might require analysis. It finds limited use as a carbide stabilizer in austenitic stainless steels, where the same remarks apply as for niobium. It can be used as an alternative to tungsten in tool steel, but is much more expensive. The investigator might require to identify whether or not tantalum was present, and if so in what percentage.

32. Tellurium (Te)

Exactly the same comments apply to tellurium as to selenium.

33. Tin (Sn)

Tin is now a relatively scarce element and therefore expensive. It is occasionally found in steels, generally as a tramp element. More information is given in Chapter 13 on Materials information.

Tin at one time was used as an electrodeposit, but this is now quite rare. There could, however, be occasions when an electrodeposit had to be identified. This would be quantitative.

There would require to be a very unusual reason why a failure investigator would require analysis for tin with steel failures.

There are some aluminium alloys which have small quantities of tin specified and the specialist investigator might require analysis.

In the copper alloys the word 'bronze' at one time meant only tin-copper. There are still a number of alloys which are tin bronzes, having between 1 and 10% tin, and depending on the mechanism of failure, the investigator may require chemical analysis.

Common solders have tin as their major constituent and the quality of soldering will be affected by the ratio of tin and lead. Depending on the problem, it would be quite common for an investigator to require analysis for tin in solder compounds.

34. Titanium (Ti)

Titanium is one of the 13 metallic elements considered to make up the majority of metals and alloys. At no time would an investigator require chemical analysis as part of a failure enquiry into titanium or any of its alloys.

Titanium is often present in many metals and alloys as a residual from the smelting operation as it is a very common element on the Earth's surface in the form of titanium dioxide.

Like oxygen, sulphur, silicon, etc., titanium can be present in the slag or oxide content of the iron or steel. As with these elements, the investigative metallurgist is unlikely to require to know the exact proportion of each where a failure is being investigated.

With austenitic stainless steels, titanium is added to prevent 'weld decay' in some alloys. The problem can also be prevented using niobium (Nb) (columbium Cb), or by reducing the carbon content.

There is a test – the Strauss test – which can predict 'weld decay' and an investigator would almost certainly prefer this to chemical analysis. Details are given under niobium in this section.

Some duplex stainless steels have titanium added and like other precipitation hardening elements it can be more economical to carry out chemical analysis than heat treatment tests. The same comments apply to the age hardening of nickel chromium alloys.

Titanium carbides are sometimes used for either refractory or abrasion purposes. In most cases the quantity, distribution and particle size will be more relevant than whether it is titanium carbide or some other carbide present. Thus analysis of these materials would not normally be required.

Titanium oxide is commonly used as a white paint pigment, but it would be unusual for analysis to be required. Titanium dioxide can also be used as a filler or 'extender' in plastics/polymers but would not normally be subjected to analysis.

35. Tungsten (W)

This finds some use because of its high density – specific gravity – but would never be subjected to analysis.

It is a common alloy element in many steels, particularly tool steels. The metallurgist investigating a failure would see the distribution of carbides, but might require analysis in certain circumstances.

Tungsten carbides are common cutting tools, but it would be seldom that analysis for tungsten would be required. There are copper tungsten alloys used for electrodes where on occasion analysis could be required.

36. Vanadium (V)

This element is found with many heavy duty alloy steels, particularly where toughness and high tensile strength are required. The investigator will be aware from the mechanism of failure whether or not vanadium could have had an influence on the failure. If, for example, a spring had failed in a brittle manner with no evidence of any stress raisers or overload and was supposed to contain vanadium, then the investigator would be justified in requesting vanadium analysis.

Where hot corrosion on nickel alloy components in gas turbines is found, vanadium might be involved. The ratio of this element to others in the corrosion debris can supply information to the specialist.

37. Zinc (Zn)

This is one of the 13 metals used in industry. Zinc components tend to be at the lower end of the economic market and thus failure investigation is seldom considered for economic reasons.

It can be difficult to differentiate visually between zinc and aluminium; thus the investigator might require to know the material involved. This would only require qualitative analysis.

Zinc is used on a large scale as an electrodeposit and in hot dip galvanizing and sheradizing for the protection of iron and steel components.

Where corrosion or contamination has been identified, therefore, chemical analysis of a qualitative nature might be justified to identify whether or not the component had been zinc coated. The difference between zinc and cadmium cannot be identified by visual examination and thus again there could be circumstances where analysis of the coating would be required. The investigator should be aware, however, that if the coating is specified as zinc and in fact has been cadmium plated, this would give better corrosion resistance and thus should not have taken part in any investigation.

The same comments apply to aluminium.

There might, however, because of Health and Safety requirements, be a need to check that the coating was indeed zinc and not cadmium, which has a high health risk.

There are some aluminium alloys with zinc present. These are precipitation ageing alloys and thus the investigator handling tensile problems would be justified in carrying out chemical analysis rather than expensive heat treatment experiments.

In the copper series, the majority of alloys at one time came under the heading of brass, which is a copper and zinc alloy. There are two brasses, alpha and beta, the difference being the quantity of zinc relative to copper. The investigator might be able to identify from the mechanism of failure and the specification whether or not the material is of the correct type. Analysis for zinc could be required to prove the type involved.

With non-metallic materials zinc will seldom or never be required for analysis.

38. Zirconium (Zr)

This metal is sometimes used in the nuclear industry, where any failure investigation is carried out by a specialist and is outside the remit of this book.

Zirconium is found in steels firstly as a residual from the ore present in the oxide or slag, and secondly as a deoxidizing agent to combine and fix the oxygen.

Zirconium oxide is readily identified by the metallurgist by its particle shape and copper colour and thus analysis for zirconium should never be necessary.

Electron microscopes

12

The optical microscope makes use of reflected light to magnify up to ×1000 to examine metallurgical structure. In the mid-1920s it was discovered that using electrons it was possible to magnify and examine metallic structures at higher magnification.

RESOLVING POWER

Development work was carried out and in the 1950s the first practical electron microscope was produced, capable of magnifying structures up to ×10,000. Since then advances in the technique have increased the magnification, the modern electron microscope being capable of magnification up to ×250,000.

A feature of electron microscopes is not just their resolving power but their ability to recognize most if not all elements present in a sample.

The metallurgical investigator does not require to be an optician, or an expect in optics in order to identify the structures which can be seen at magnifications up to ×1000. Likewise the investigator does not require to be an electronic wizard to obtain information from the electron microscope. The information is quite distinct, however, from optical images, and must therefore be interpreted by a trained observer. Similarly, it is not really possible to appreciate the significance of the structures by self-education, thus training and experience are essential.

Common sense dictates that any component which does not have a problem identifiable at less than ×1000 magnification would not fail under normal circumstances. A component for which the reason for failure was a defect only identifiable by electron microscopy would have a fundamental material design problem which would seldom be presented for normal investigation.

Two types of electron microscope are now available, both operating under high vacuum:

- (TEM) Transmission electron microscope
- (SEM) Scanning electron microscope

TRANSMISSION ELECTRON MICROSCOPE (TEM)

This form of electron microscope is used for research and development. A beam of electrons is fired through the specimen – usually a replica of the specimen surface – and the image produced by transmitted, reflected or secondary electrons is then viewed on a monitor screen. It must be noted that the specimen presented to the TEM must be extremely thin to allow the electron beam to pass through.

SCANNING ELECTRON MICROSCOPE (SEM)

This form of electron microscope can be used in failure investigation as well as research.

As the name suggests, an electron beam is fired and 'scanned' over the specimen under computer control. The image produced has an almost three-dimensional appearance and requires training and experience for interpretation.

The specimen can be a standard micro-specimen slightly modified, the surface of a fracture, or a component surface. On occasion a damaged fracture face can supply information not readily available at the visual examination.

All samples used in the SEM can be polished and etched in the same manner as normal micro-specimens for the optical microscope, although etchants, which rely on staining rather than chemical attack, cannot be used as the image produced would be poor.

One potential problem with examination of untreated fracture faces is that if the specimen has been manhandled, peculiar results can be obtained, possibly misleading the investigator, particularly where corrosion is suspected which masks the underlying structure and cause of failure.

The SEM has several variations almost always known by acronyms, some of which are briefly described below. It is rare that any are involved in failure investigation.

Energy dispersive X-ray analysis (EDAX)

This is a form of electron probe micro-analysis. It is used for identifying any metals present in the prepared specimen visually and chemically. It works on the basis that an element has certain electron absorption characteristics and can distinguish between different metals present based on the electron energy which is transmitted (absorbed) or reflected. It can identify the presence of any known element, and give a reasonably accurate estimate of the amount present.

Electron Spectroscopy for Chemical Analysis (ESCA)

This is a general term given to any chemical analysis technique used on the SEM.

Secondary Ion Mass Spectrometry (SIMS)

This uses several techniques for ion mass spectrometry based on secondary emission from the surface of specimens bombarded by electrons. It is a useful tool for analysis of surfaces for metallic, organic or inorganic contamination or coatings. The sample needs little preparation and can supply information for sophisticated corrosion or contamination problems. These are unlikely to be met with every day, but there could be occasions where the investigator requires further informaton and this technique might then be of value.

THE VALUE OF ELECTRON MICROSCOPES

From the above it will be seen that the electron microscope can be useful when additional information is required following the normal visual, metallurgical, macro/micro-examination and hardness and mechanical testing.

In some industries such as the electronics industry, where very thin surface coatings are used, it is an essential tool in production control and failure analysis.

There is, however, evidence that in some cases the laboratory with an electron microscope uses it at an early stage of an investigation, and that photomicrographs at $\times 2000$ plus are produced with information supplied which in many cases results in confusion.

There is no doubt that the electron microscope can confirm and enlarge on the evidence identified by the metallurgist at the visual and metallurgical examinations using an optical microscope. Whether or not this confirmation is justified by the high cost of electron microscope examination has to be judged on an individual basis.

There are a number of case histories presented in this book, none of which required the use of electron microscopy for the basic information, and only one discusses how the initial information was refined from a 'copper alloy' to an 'aluminium bronze', using SEM. This is typical of the author's experience in over 40 years of failure investigation.

It is relevant that the author has been employed on several occasions to interpret metallurgical reports which used the electronic microscope.

PART FOUR

Reference Section

Materials Information 13

In failure analysis, the investigator must be aware of the properties of materials, and the relevance of these properties to the cause of failure. This chapter describes in detail the 13 elements considered to be the basis of all materials used in industry. It is suggested that carbon and silicon sub-divide into five groups, giving 16 groupings in all.

It must be appreciated that this classification is related to the author's opinion and practical experience, and will lack the detailed knowledge of any specialist: the choice of groupings will also differ between individuals.

Each group chosen has a brief introduction on its history and any relevant interesting information, with common names, and UK and USA specifications. Any investigator should be aware of the use for which the material was intended, as misuse could be a major reason for failure. Some information on the possible mechanism or reason for failure is supplied, but this must of necessity be quite generalized.

The technique or techniques which might be involved in a failure investigation are discussed and evaluated. There will be a measure of duplication between this information and that given in Chapter 10 on Mechanical testing and Chapter 11 on Chemical analysis. Table 13.1 lists the 13 elements with their major alloys and compounds, giving approx. 90 materials in all, placed in 16 groupings.

A IRON (Fe) AND ITS ALLOYS

This is by far the most important element in metallurgy and engineering. It bears the latin name *ferrium*, hence the symbol Fe.

Table 13.1 Index to elements and their compounds

A	Iron (Fe)
A1	Wrought iron
A2	Cast iron
A2a	Grey cast iron
A2b	Spheroidal graphite (SG) cast iron – ductile iron – Mechanite
A2c	White cast iron
A3	Iron–nickel alloys
A4	Steel
A4a	Mild steel 0.05–0.25% carbon – low carbon
A4b	Mild steel 0.20–0.5% carbon – medium carbon
A4c	High carbon steel – 0.4–1.2% carbon (Stubbs) steel
A4d	Low carbon alloy steel 0.1–0.3% carbon, 5% alloy max.
A4e	Medium carbon alloy steel 0.25–0.6% carbon, 5% alloy max.
A4f	High carbon alloy steel 0.5–1.2% carbon, 5% alloy max.
A4g	Medium carbon tool steel 0.2–0.6% carbon, 5–20% alloy
A4h	High carbon tool steel 0.5–1.2% carbon, 5–20% alloy
A5	12% chromium stainless steels
A5a	Low/medium carbon 12% chromium steels
A5b	High carbon 12% chromium steels
A6	18/8 austenitic stainless steels
A6a	18/8 chromium–nickel austenitic stainless steels
A6b	18/8 chromium–nickel austenitic stainless steels – weld stabilized
A6c	18/8 chromium–nickel austenitic stainless steels – 2–3% molybdenum
A6d	High chromium–nickel (up to 20/20) austenitic steels
A7	Duplex stainless steels
B	Aluminium (Al)
B1	99% aluminium
B2	1–2% manganese–aluminium alloys
B3	Magnesium–aluminium alloys
B4	Silicon–aluminium alloys
B5	Copper–aluminium alloys
B6	Magnesium-silicon–aluminium alloys
B7	Zinc–aluminium alloys
C	Copper (Cu)
C1	99% copper
C1a	99% electrolytically refined copper
C1b	Phosphorus/arsenical/fire–refined copper
C1c	Low alloy coppers – up to 1% alloy content
C2	Brass, zinc–copper alloys
C2a	Alpha brass – 40/60 zinc–copper
C2b	Beta brass – 30/70 zinc–copper
C3	Bronze, tin–copper alloys
C4	Cupro-nickel: nickel–copper alloys
C5	Nickel-silver: nickel–zinc–copper alloys
C6	Aluminium bronze: aluminium–iron–nickel–copper alloys
C7	Beryllium–copper alloys
C8	Chromium–copper alloys
C9	Copper brazing alloys

D	Zinc (Zn)
D1	99% zinc
D2	Zinc alloys
E	Nickel (Ni)
E1	99% nickel
E2	20% chromium–nickel alloys – non-age hardening
E3	20% chromium–nickel alloys – age hardening
E4	Copper–nickel alloys
E5	Iron–nickel alloys
E6	Beryllium–nickel alloys
F	Titanium (Ti)
F1	99% titanium
F2	Titanium alloys – non-heat-treatable
F3	Titanium alloys – heat-treatable
G	Cobalt (Co)
G1	99% cobalt
G2	Cobalt alloys
H	Magnesium (Mg)
H1	Magnesium alloys
I	Tin (Sn)
I1	Tin and tin alloys
J	Lead (Pb)
J1	Lead alloys
K	Tungsten (W)
K1	Tungsten alloys
L	Silicates – silicon-based
L1	Stone and aggregates
L2	Concrete, cement, mortar, bricks
L2a	Cement/concrete
L2b	Bricks, tiles
M	Silicates – silicon-based man-made materials
M1	Glass
M2	Ceramics
M3	Refractories
N	Carbon products
N1	Wood
N2	Wood products
O	Plastics/polymers – carbon-based
O1	Thermoplastic materials
O2	Thermosetting materials
P	Natural materials – carbon-based, fibres or sheet materials

There are seven headings or common names for 'iron alloys' or compounds used in this book:

- A1 Wrought iron
- A2 Cast iron
- A3 Iron–nickel alloys
- A4 Steel – plain carbon and alloy steels – up to 20% alloy content
- A5 12% chromium stainless steels
- A6 18/8 chromium nickel stainless steels (austenitic)
- A7 Duplex stainless steels

A1 Wrought Iron

This allowed the industrial revolution to proceed rapidly. It contains no appreciable carbon but can have considerable quantities of oxide in the form of iron oxide and silicon oxide (silica), and phosphorus and sulphur as iron phosphide and iron sulphide. There could also be aluminium as aluminium oxide (alumina). Wrought irons with up to 25% impurities or non-metallic inclusions are not uncommon.

Swedish Iron and Best Yorkshire are common names for wrought iron. BS 48, BS 51, BS 858 and ASTM A42 are national specifications for wrought iron.

Pure iron has a high ductility and excellent corrosion resistance, and wrought iron is still in use throughout the world, although it has not been produced commercially for the best part of 40 years. Wrought iron with low levels of impurities, less than 2%, can be expected to have excellent ductility, the ability to accept cold work without excessive hardening, and excellent corrosion resistance. (The Delhi Pillar in India is a corrosion-free lump of iron existing in high humid conditions. It is pure iron, believed to be a meteorite.)

Wrought iron was used for items such as bars, plates, girders, lifting chain, ships' anchors and chain, and any product where strength was required with ductility. Ship and boiler plate, rails, structural beams and wire are examples of the products produced. Figure 13.1 shows typical cross and longitudinal sections.

When the non-metallic, that is the oxide and slag content increases, the impact and to some extent the ductility measured as elongation will be reduced, sometimes quite dramatically. The through thickness properties in particular can be very poor. This can be like a book which is quite strong when pulled vertically or horizontally, but very weak when it is opened. This weakness can be seen in the micro-structure. Failure could be brittle, resulting from carbon pick-up, excessive grain size or oxide. Where fillet welds are attached to a wrought iron surface this can apply a tensile load which could 'open the book', so to speak.

There are very few occasions when chemical analysis of wrought iron would supply any meaningful information to the failure investigator. The microstructure will show stringers of non-metallic material, generally grey or black in colour, and elongated where they have been worked.

Mechanical testing might be necessary to identify the yield, ductility and impact properties.

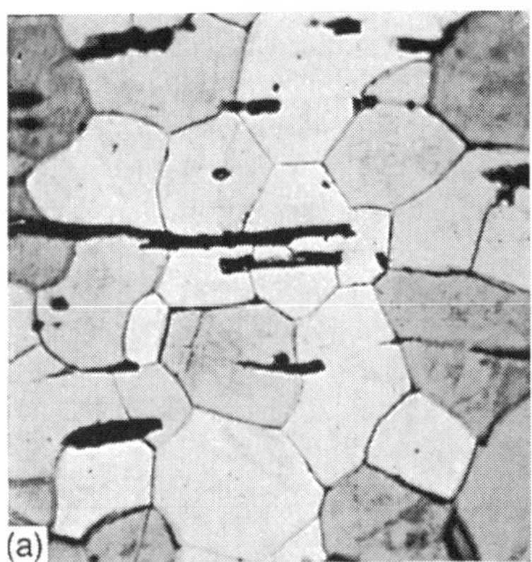

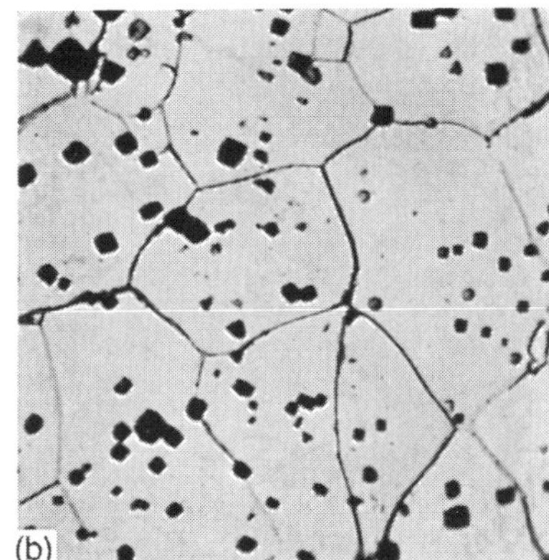

Figure 13.1 Photomicrographs showing (a) a longitudinal and (b) a cross-sectional view of wrought iron at ×250 magnification.

A2 Cast iron

This has iron plus carbon (graphite) with or without iron carbide (cementite) in the cast state. It cannot be hot or cold worked, and exists in three forms which are illustrated in Figure 13.2.

Grey cast iron. Here the majority of the carbon is present as flakes of graphite. The amount of carbon present as iron carbide will vary from zero to 0.8%.

Spheroidal graphite cast iron. Either in the form of Meehanite, where the flakes are small in size, or the more common spheroidal graphite (SG), where the graphite is in the form of spheres. Again the combined carbon will vary from zero to 0.8%.

White cast iron. There will always be at least 0.8% carbon present as iron carbide.

A2a Grey cast iron

There will be some considerable variation in the metallurgical structure and mechanical properties. National specifications are BS 1452, ASTM A278.

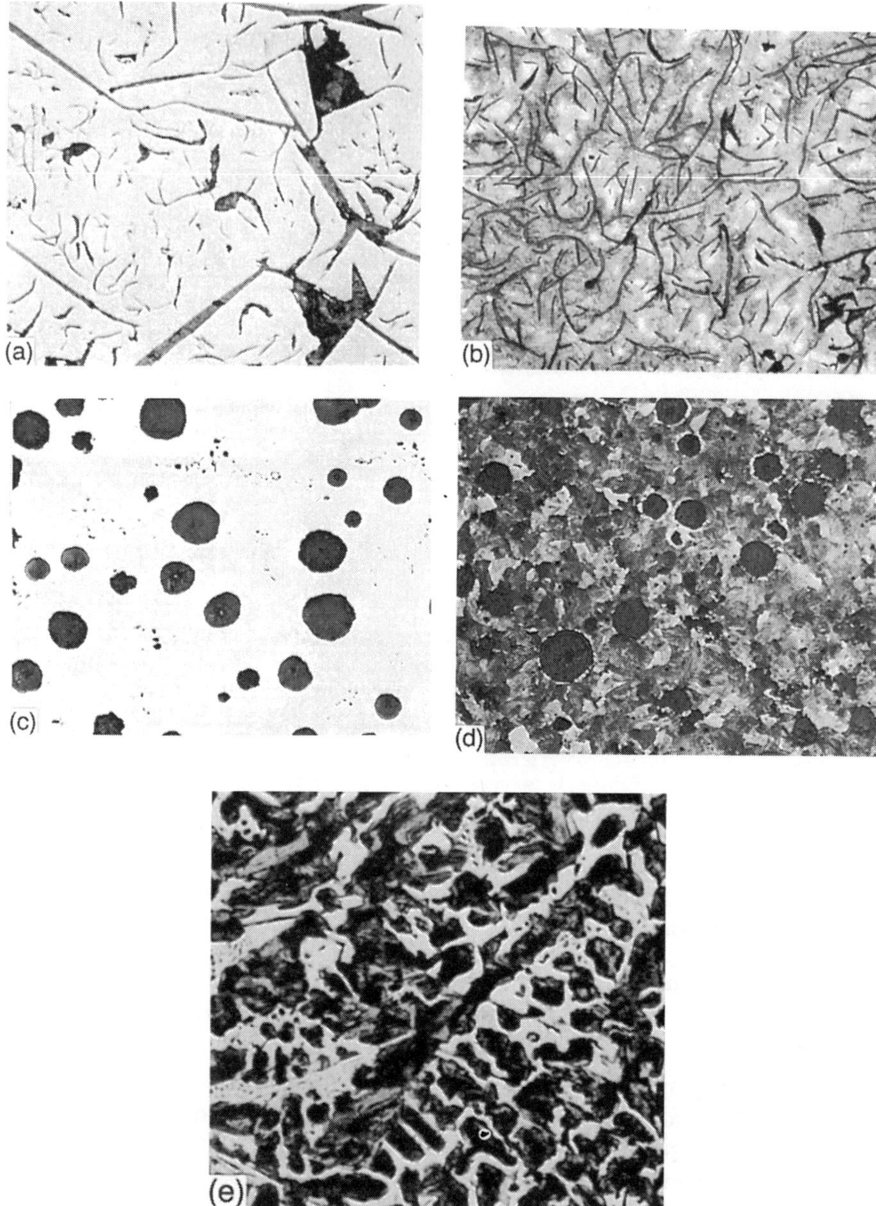

Figure 13.2 Photomicrographs of cast iron. (a) grey cast iron at ×100, fully ferritic, all carbon as graphite; (b) grey cast iron at ×100, pearlite, 0.7% of carbon as iron carbide, remainder graphite; (c) spheroidal graphite (ductile), fully ferritic cast iron at ×150, all carbon present as graphite; (d) spheroidal graphite (ductile), fully pearlitic cast iron at ×150, 0.7% of carbon present as iron carbide, remainder graphite; (e) white cast iron at ×150, no graphite present, 0.6% carbon as iron carbide in pearlite, remainder as graphite.

Grey cast irons are used for engineering components not subjected to impact, bending or high tensile loading. Intricate shapes can be economically produced, with high phosphorus irons giving excellent detail but very poor impact properties. Grey cast irons are still used for castings of all types, and frames, or 'beds' for many types of machines. They are readily machinable, have considerable stability, and can absorb vibrations, thus reducing noise. With the cast skin intact they have excellent corrosion resistance.

The majority of grey cast iron failures are brittle tensile failures from porosity, coarse graphite, surface decarburization, surface chilling or damage. These materials will not accept bending, impact or tensile loading with or without these defects. Welding, unless very carefully controlled, should not be attempted.

Many low cost items previously produced in grey cast iron are now made in plastic/polymer, aluminium or zinc castings, or mild steel welded fabrications.

A2b Meehanite and SG iron

These have the graphite inhibited either by controlling the size of the flake, as with Meehanite, or ensuring the graphite is present as balls, as with SG (spheroidal graphite) cast iron. BS 2789 and ASTM A220 are the national specifications. Ductile iron is a common name.

These are used for many engineering components such as crankshafts, camshafts, gears, etc. They have much better ductility and impact resistance, but lower machinability than grey iron, and less ability to absorb vibration.

Failure could be fatigue from surface effects, porosity, presence of flake graphite, surface decarburization, surface chilling or welding. Welding of any type of cast iron can be expected to result in poor impact properties, tensile strength and ductility. This can be readily identified by micro-examination.

A2c White cast iron

These have most, if not all, the carbon present in the form of iron carbide with little or no free graphite.

White cast iron finds limited use where high abrasion resistance is required which will not be subject to shock or tensile stress. It is glass hard, but like glass is readily shattered.

With any of the cast irons, the investigator should be able to tell by visual examination the mechanism of failure. The micro-structure and hardness should then give the reason for failure and it would be very rare that chemical analysis could contribute more to any cast iron failure investigation.

The metallurgist can identify the graphite/pearlite ratio, oxide or slag, grain size and phosphide content. The type and distribution of graphite will be identified. Casting defects such as shrinkage and porosity are readily seen, with problems such as hot or cold 'shuts' recognized. These are where the molten metal has not diffused, generally because of a surface layer of oxide.

This information is more valuable and meaningful than any chemical analysis can supply. Tensile tests can be useful, but impact testing will be called for only with SG and Meehanite irons, as grey or white irons cannot be expected to exhibit any significant impact strength. Additional information will be found in Chapter 5 on Micro-examination.

A3 Iron–nickel alloys

There are relatively few specifications for these materials which have significant quantities of nickel, very often in the 50% Ni, 50% Fe range. They exist for their magnetic/non-magnetic properties, and to a lesser extent for low thermal or controlled expansion characteristics. Failure investigation, if related to these physical characteristics, would require chemical analysis.

If, however, the failure was mechanical, then the mechanism and metallurgical condition would be of much greater relevance than the chemical make-up.

A4 Steel

Steel is an alloy of iron and iron carbide invariably with other elements present. It is not an alloy of iron and carbon but the percentage of carbon present obviously decides the amount of iron carbide.

Depending on the method of classification there can be over 20 groups of plain carbon and low alloy steels, with separate groups under the heading of stainless steel.

The plain carbon and low alloy steels are here divided into eight categories with their properties and uses described, plus information on the specifications which would be involved. This is the author's classification and it is accepted that there will be other systems or groupings. In addition there are 5 categories for stainless steels where there is likely to be less dissention.

Chapter 14 on Heat treatment discusses how steel properties can be dramatically altered by controlled and uncontrolled heating and cooling.

A list of the eight categories of plain carbon and low alloy steels is:

A4a	Mild steel	0.05–0.25% max carbon steel, alloy content less than 1%
A4b	Medium carbon steel	0.20–0.55% max carbon steel, alloy content less than 1%
A4c	High carbon steel	0.5–1.0% max carbon steel, alloy content less than 1% – Stubbs steel

A4d	Low carbon alloy steel	0.1–0.3% carbon, high carbon steel 5% maximum
A4e	Medium carbon alloy steel	0.25%–0.55% carbon steel, alloy content 5% maximum
A4f	High carbon alloy steel	0.5–1.0% carbon steel, 5% maximum alloy content
A4g	Medium carbon tool steel	0.2–0.5% carbon steel, 5–20% alloy content
A4h	High carbon tool steel	0.45–1.1% carbon steel, 5–20% alloy tool content

The following text discusses each of the above listed groups.

The modern idea of specifying the carbon equivalent (CE) can be very useful, particularly where welding or heat treatment is involved. This gives an evaluation of the hardenability of a steel. The calculation is:

$$CE\% = C\% + \frac{Mn\%}{6} + \frac{Cr\% + Mo\% + V\%}{5} + \frac{Ni\% + Cu\%}{15}$$

There are some variations on this, but the above is the British Standard.

A steel with a CE below 0.4–0.45% would be a reasonable definition of mild steel. This will have low hardenability, thus would need little or no preheating at welding, and any section thicker than 5–10 mm could not be efficiently quench hardened. More information on this is given in Chapter 14 on Heat treatment.

It must always be appreciated that when metallurgists are involved there will be some divergence of opinion on the meaning and the implication of technical terms, and this certainly applies to 'hardenability'.

A4a Mild steel

0.05–0.25% maximum carbon with maximum alloy content of 1.0%. CE% max. 0.4–0.45%. Embraces mild steel and low carbon steel to BS 4360-43, BS 970 En2, BS 970 040 A10, AISI 1020.

These make up a high proportion of steels used by weight or volume. The majority will be in the form of bar, strip and sheet with some forgings and castings.

It should be noted that the term 'mild steel' has quite different meanings in different areas and industries. The bodywork of motor cars, ship plate, most engineering casings and considerable quantities of structural steel come within the category of mild steel. 'Tin cans' were made of this grade. Some steel in this group can be carburized but will be for low duty components. The higher carbon steels in the section can be flame or induction hardened to a limited extent.

Competition from stainless steel, aluminium, zinc die-castings and also plastics/polymers now exists.

The steels have a carbon content of 0.3% maximum, generally less than 0.25%. The materials have low mechanical strength and can be expected to have variable degrees of ductility and impact resistance. The carbon equivalent (CE) will always be less than 0.45%.

It is unusual for materials in this group to be used where tensile strength is required. Most problems are found with brittle failure or fatigue failure resulting from welding, heat treatment or impact at low temperatures. Micro-examination with hardness surveys will often identify the cause of failure. Tensile and impact tests on other occasions will be necessary.

Corrosion is the main problem with these materials. This is discussed in Chapter 7 on Mechanisms of failure. Stress corrosion, however, is unusual but can exist. Pitting corrosion (crevice) is common, as is chemical and galvanic corrosion. Electrolytic corrosion from stray currents is probably more frequent than is normally appreciated.

A4b Medium carbon steel

0.20–0.55% carbon maximum, alloy content of 1.0% maximum. CE would be 0.35% min – 0.6% max.

Standards BS 4360-50, BS 970 En8, BS 970 080 M40, AISI 1040 apply.

These make up the bulk of structural steels, shafts, springs (low duty), plate, casings, gears, etc. of medium low tensile strength.

Certain of the lower carbon steels can be carburized. Most of these steels can be successfully flame or induction hardened.

These steels would be expected to have a carbon equivalent (CE) of between 0.35% and 0.6% and thus should be pre-heated at welding. The pre-heating would be in the region of 75–120°C.

Failure can be by impact, fatigue, tensile (ductile or brittle) stress, shear, abrasion or corrosion. Corrosion is a major problem with this group, which is susceptible to stress corrosion, brittle failure from corrosion pits, and fatigue initiated from corrosion damage. Welding of the higher carbon steels requires pre-heating and careful control at welding. Weld problems are a common cause of failure.

A4c High carbon steel

0.5–1.0% carbon maximum, alloy content 1% maximum, includes 'Stubbs steel'. Specifications BS 24/3 A7, BS 970 En43A, BS 970 080 M50, AISI 1080 refer.

These steels almost always have a carbon equivalent above 0.7% and thus good hardenability and are generally used in the hardened and tempered condition. They are seldom welded.

Medium duty springs, punches, chisels, spring washers, low duty cutting tools, etc. are made from this group. They will have high hardness/abrasion resistance, with low ductility.

Tempering or heating above approximately 300°C will result in the higher hardness components being softened, this can be either by thermal or friction.

Failure can be expected to have a brittle connotation. Ductile failure could indicate the wrong material, or faulty heat treatment. Fatigue failure from stress raisers, corrosion pits, damage marks, etc. can be expected.

Tack welds, weld spatter or stray arcing is a common reason for brittle failure of these steels. Corrosion and hydrogen embrittlement at electroplating also results in serious problems. Welding requires a high degree of control, and in many organizations would be banned. Minute stress corrosion cracking can result in brittle failure and can be difficult to identify in the failed component.

A4d Low carbon alloy steel

0.1–0.3% carbon with maximum 5% alloy content. CE = 0.6% min. – 0.85% max. Specifications BS 970 En357, BS 970 En36, BS 970 655 M13, BS 970 653 M15, AISI 8637 refer.

This group includes heavy duty carburizing and some nitriding steels, along with medium strength, high ductility steels for bolts, studs, connecting rods, drive shafts, medium to high tensile wire and rod components, medium duty springs, crankshafts, etc. Induction and flame hardening requires special control.

The alloying elements in the main are chromium, nickel and molybdenum, with vanadium, tungsten and others to a lesser extent.

This group has no better corrosion resistance than carbon steels. Stress and pitting corrosion can have serious results from fatigue initiation or brittle fracture. Minute stress corrosion cracks act as stress raisers for fatigue propagation, and can be difficult to identify.

Any welding process will require pre-heating and control of heat input. Electroplating of the higher carbon, higher alloy steels could result in hydrogen embrittlement unless stress is relieved immediately after plating.

A4e Medium carbon alloy steel

0.25–0.55% carbon with 5% maximum alloy steel. CE = 0.65% min. – 1.0% max. Standards BS 970 En19–En24, BS 970 708 M40, 817 M40, AISI 4140, AISI 4340 apply.

This group makes up the bulk of the heavy duty, high strength components used by engineers, such as crankshafts, drive shafts, many springs and high strength case hardening steels. It includes carburizing, nitriding and carbonitride steels, but these are not suitable for induction or flame hardening without special control. There will be some cutting tool steels in this group, also many general-purpose steels used for high strength components.

They all rely on high quality heat treatment, with clean steel free of surface defects.

Failures will include embrittlement, torsion, shear and fatigue failure. The steels in this group are susceptible to stress raisers to initiate fatigue. The investigator will require to identify if stress raiser damage, faulty heat treatment, or wrong material grade is the problem. Hydrogen embrittlement failure can be expected if the higher carbon materials are electroplated. Stress-corrosion cracking propagating by fatigue is not uncommon.

Welding should only be carried out with weld procedures prepared and correctly applied. Stress release immediately following electroplating will be essential.

The alloy elements will include chromium, nickel and molybdenum, with vanadium, tungsten, cobalt and others. These steels have no better resistance to corrosion than plain carbon steel but the significance of pitting and stress-corrosion cracking is much more serious. Welding requires careful pre-heating and control of heat input.

Chemical analysis is commonly required by the investigator.

A4f High carbon alloy steel

0.5–1.0% carbon with 5% maximum alloy content. CE = 1.0% upwards. Relevant specifications include BS 970 En31, ASTM A485, SAE 8660, F1, F3 tool steels.

These steels are used for heavy duty springs, drive shafts and similar components. Cutting tools will be in this group. Ball and roller bearings are normally made from 1% carbon, 1% chromium steels.

These steels are all very susceptible to stress raisers initiating fatigue and have no better corrosion resistance than plain carbon steels. Corrosion pitting and stress-corrosion cracking can result in fatigue initiation or brittle failure. Failure will generally be brittle. They are very susceptible to hydrogen embrittlement and should not be electroplated, except under carefully controlled conditions with stress relieving prior to and after plating.

Stress-corrosion cracking of this group, initiating fatigue or brittle failure is probably more common than is generally appreciated. There is a case history reported where stress-corrosion cracking of a minute degree initiated fatigue failure. Heat treatment can be critical and must be carefully controlled.

Welding of these materials is not advised, and certainly technical advice is essential regarding pre-heating, heat input and post-weld stress release. Many organizations ban all welding.

A4g Medium carbon tool steel

0.2–0.5% carbon, 5–20% alloy content. The carbon equivalent (CE) is always high and thus not a variable factor.

These steels have significant quantities of chromium (max 8%), molybdenum, vanadium, cobalt, titanium, etc. Steels with more than 10% chrome will be in the stainless steel group A5. The steels in group A4g are used for cutting, abrasion resistance, and high tensile purposes. They have inherently poor ductility and impact resistance, and in most cases little corrosion resistance.

Failure can be expected to be brittle. These steels should not be welded or electroplated without metallurgical advice and great care is required to eliminate all sharp radii and other stress raisers. They would not be expected to withstand impact or bending.

The heat treatment control in most cases will be of more significance than chemical analysis. Where the metallurgist finds large grains or peculiar metallurgical structures, this will supply more information to the investigator than would chemical analysis.

A4h High carbon tool steel

0.45–1.1% carbon, 5–20% alloy content. The carbon equivalent of these steels would not normally be considered to be significant but it will always be high.

These are principally tool steels for cutting, milling, broaching, etc. There is little history of these steels having failures investigated. The users are aware that they are extremely brittle and thus cannot stand abuse. They tend to be expensive and are used by skilled personnel. They should not be electroplated or welded except under exceptional circumstances with practical metallurgical advice.

A5 12% chromium stainless steels

These include AISI 400 series, BS 970 400 series, BS 970 En56 series steels.

General information

These stainless steels, which are also known as 12% chrome steels, contain a minimum of 9–10% chromium with a limited amount of other alloy elements and variable carbon. They are magnetic steels and can be subjected to heat treatment processes standard for alloy steels. They are generally hardened by quenching and tempering, sometimes nitriding, but not carburizing.

Their corrosion resistance is achieved by the adherent, coherent chromium oxide which forms immediately and is therefore self-healing in an oxidizing atmosphere. These steels therefore differ from the austenitic steels, type A6, which, in addition to having the self-healing chromium oxide film, are single-phase materials and have significantly better corrosion resistance than the 12% chromium steels. The 12% chromium steels, however, have the very considerable advantage that they can be hardened and tempered, and thus do not suffer from the low mechanical strength which is characteristic of austenitic stainless steels.

Because these steels are used for their mechanical properties in addition to corrosion resistance, it can be expected that a high proportion of the failures investigated will be the result of mechanical failure and all the comments which apply to the examination of alloy steel failures apply to these. That is, failure can be expected to be due to fatigue, particularly initiating from crevice and stress corrosion. Fatigue from other sources, such as surface imperfections, slag inclusions, etc., is not uncommon.

Corrosion failures can be expected from denuding of the chromium at the surface by heat treatment, welding or some local corrosion. It must be appreciated that these steels in the correct condition do not rust, that is form brown/red ferric oxide on the surface. Their corrosion products are greenish blue or black. Very often the corrosion takes the form of bright pits – crevice corrosion. This is the result of covering the surface and then abrading the covered surface. The absence of air/oxygen prevents the self-healing reoxidation process from occurring and if there is a corrosive environment the surface will corrode in the form of pits, which can be bright and are easily recognized as crevice corrosion. These crevices/pits can be of substantial depth and may penetrate tank walls, causing leakage.

There is evidence that when the steels are nitrided the corrosion resistance of the nitrided layer will be dramatically reduced. This can result in very hard abrasive particles which destroy the evidence of corrosion. There is only limited information available at present in this area, but sea-water control valves and plastic moulds have given problems of this nature.

Some specifications are formulated to give crevice and stress corrosion protection, that is they have a more adherent, corrosion-resistant oxide film.

Some of the steels have been developed to the extent where they are no longer 12% chromium and are now within the duplex group of stainless steels, A7.

Under some conditions it is possible for these steels to corrode by rusting. When the chromium content is reduced below 9–10%, the self-healing chromium oxide does not readily form; thus the surface can rust. Very often, however, the rust found is the result of surface contamination.

The ferroxil test described in the section on austenitic steels (A6) and in that on Corrosion in Chapter 7 can be used to assess whether or not the steels are 12% chromium. If the surface is rusting then it should give a positive result to the ferroxil test, that is it should turn a deep blue. If the surface is then lightly abraded with clean non-metallic grit and re-tested, the test is negative if the material has no free iron, and if the material is magnetic, it can be expected to be of the 12% chromium type.

A5a Low/medium carbon 12% chromium steels

These have 0.05–0.3% carbon, and up to 2% alloy content. They are covered by specifications BS 970 En56B, BS 970 En57, BS 970 400/S29 and AISI 400.

The very low carbon steels in this range are known as chromium irons as the low carbon content precludes the formation of any significant iron carbide. As these are almost single phase materials with 12% chromium producing the adherent oxide, they have excellent corrosion resistance, are easily formed and are used on many occasions on components where a reasonable level of corrosion resistance is essential. Items such as wheel hubs and certain parts of car trims are sometimes made in these materials. They have low strength but excellent ductility, and are cheaper than austenitic steel. Household items such as pots, pans, forks and spoons are made with these steels.

Failure of these steels could be due to rusting caused by denuding of the chromium at the surface, or a result of misuse. They have low strength, and ductile failure could be expected if they are overloaded. Brittle failure would require investigation.

The higher carbon steels in this group are use for components of medium strength and corrosion resistance.

In some areas they are competing with the 18/8 austenitic steels because they can be cheaper, but normally they are chosen because of their higher mechanical strength in corrosive conditions.

Many components in the petrochemical, food, brewing and distilling industries use these materials for their mechanical strength and corrosion resistance. Gas turbines use them for components such as compressor blades, and in some cases turbine blades. The same components will be found in the higher carbon group, A5b.

Comments regarding failure mechanisms, etc., given in section A5b also apply to these components.

With the addition of 1% nickel, steels of better ductility and slightly better corrosion resistance are obtained. These higher carbon steels are used for aircraft engine components.

The addition of molybdenum will stabilize the chromium oxide, giving slightly better corrosion resistance.

A5b High carbon 12% chromium steels

These have 0.25–1.2% carbon and up to 4–5% other alloy elements. Specifications AISI 440C, KE970, BS 970, BS 970 En56D, BS 970 En57 refer.

These are commonly used for knives and other cutlery items in the domestic/catering market. They will be found to rust on some occasions where two components are left in contact with each other in moist conditions but the rust can be readily removed as it is only the result of reduction of the chromium content on the surface. It is important to remove this to prevent pitting.

The higher carbon steels are used for guillotine knives, punches and other such components where a sharp edge is required with high hardness and corrosion resistance. It was discovered some years ago, in particular in the razor blade industry, that while there is adequate hardness with 1% carbon steel to give a very sharp edge, this sharp edge is rapidly lost, not because of blunting but because of corrosion. The 12% chromium steels then developed give a very much longer life under the same conditions.

Where impact resistance and ductility have to be improved, alloying elements will be added, these commonly being nickel up to 1–2%, molybdenum for carbide stabilization and vanadium for grain refinement. Some steels will have tungsten added to increase the abrasive properties.

The medium carbon 12% chrome steels are used in aircraft and industrial gas turbines, sometimes modified, for compressor discs, blades and shafts. Industrial gas turbines use them for turbine blades in addition to compressor blades, all types of shafts and discs. Again this may be modified for grain refinement, creep strength, etc.

The medium carbon 12% chrome steels are finding increasing use in the chemical and food industries for items such as shafts, mixing paddles, studs, etc. There is, even so, a tendency for them to be replaced by duplex steels, particularly in the oil industry.

There is a range of high carbon 12% chrome steels, some with tungsten and other alloy elements. These can have up to 2% carbon, with similar characteristics to 1% carbon, 12% chrome. They are used for guillotine knives, where long life abrasive properties are required. With all guillotine-type knives it is the amount and distribution of the chrome and tungsten carbides which gives the cutting ability and long life. The carbides should be small to medium in size and homogeneously distributed. Too small carbides are not efficient cutters, and larger or badly distributed carbides are plucked out, giving ragged cuts.

These problems are found by metallurgical examination, not chemical analysis or mechanical tests.

With failures in gas turbines, creep must be looked for, in addition to fatigue, torque, brittle or ductile tensile properties.

Corrosion pitting is common and is the initiation of fatigue or brittle failure in many circumstances. Stress-corrosion cracking is not unusual, but can be difficult to identify, as a minute stress-corrosion crack can initiate fatigue propagation, and there is evidence that this could be more common than is appreciated at present.

As the cost of these corrosion-resistant, heat-treatable steels is reduced with increased usage, it can be expected that designers will choose them more often in lieu of standard alloy steels.

Hardening and tempering requires considerable skill to control the grain size, carbide particle size and hardness. Some information on this is given in Chapter 14 on Heat treatment.

Welding of these steels will always require great care and specialist welding rods. Pre-heating will invariably be required, and many organizations would not consider welding except under very special conditions with any carbon content above 0.6%. The carbon equivalent (CE) formula is not relevant with these steels.

Electroplating will seldom be necessary for corrosion resistance purposes but may be required for other purposes such as the use of chromium plating for low friction or abrasion resistance. Hydrogen embrittlement can be a considerable problem and stress release must be carried out immediately with great care being taken to ensure that the components are not stressed during the electroplating process, as the hydrogen embrittlement can result in fracture during plating. Most of these components would require pre-plating stress release in addition to post-plating stress release.

A6 Austenitic stainless steels

These include the AISI 300 series, BS 970 300 series, BS 970 En58 series, and are commonly called 18/8 stainless steel.

General information

These steels are based on a chemical composition of 8% nickel, 18% chromium where the critical change points in ferrite or iron are reduced from above 700–900°C to below −30°C. This means that the ferrite has changed to austenite and also that the steel is above the Curie point, where iron ceases to be magnetic. The result is the well-known fact that austenitic stainless steel is non-magnetic with good corrosion resistance, being a single phase material, making it similar to pure metals, and thus it cannot be heat treated.

Single phase materials invariably have better corrosion resistance than materials with more than one phase, for which there will always be the possibility of galvanic cell action. Alloy carbon steels always have ferrite, pearlite, cementite, iron carbide, martensite, etc., with galvanic cells existing at the surface when moisture is present.

Also, the chromium content is above the critical 9–10% level where the adherent, coherent chromium oxide is formed on the surface in place of the loosely adherent iron oxide. Thus, like 12% chromium steels, these have good corrosion protection provided they are in an oxidizing atmosphere.

The steels have low mechanical strength, particularly the yield or proof strength, which is appreciably less than for mild steel. The ultimate tensile strength is generally better than that of most mild steels and thus the material in the annealed condition has excellent ductility measured as elongation. When cold worked the surface can become brittle and readily cracks.

Because of the low strength softness, austenitic steels can be readily cold welded with quite light pressure. This leads to the problem of 'galling' discussed in the abrasion section in Chapter 7.

Austenitic steels therefore are seldom used for mechanical strength but almost invariably their corrosion resistance property is the reason for their choice. It must always be appreciated that these materials are 'corrosion resistant' and not 'non-corrosive'.

Their ability to withstand corrosion is good with oxidizing atmospheres and acids. The corrosion resistance is reduced when the materials are used in non-oxidizing conditions, and there are classic failures reported where austenitic stainless steels have been covered, for example in the form of a lap type riveted joint, where abrasion could occur between the two faces, which are not allowed to re-oxidize. When corrosive matter enters this gap, crevice corrosion or stress corrosion will almost invariably occur.

Austenitic stainless steels are also prone to stress corrosion cracking where a combination of tensile load and corrosive conditions exist. This is discussed in the corrosion section in Chapter 7. They are particularly prone to chloride attack.

These materials find their major use in the food and chemical industries, particularly brewing, distilling and petrochemicals. Most breweries, distilleries and refineries have vast quantities of austenitic stainless steel of various types – and also the marine industry, where provided the correct grade of stainless steel is chosen, there will be resistance to atmospheric corrosion. This is type 316 steel, which contains up to 3% molybdenum to stabilize the surface oxide.

There is no, or extremely little, ferrite present in austenitic stainless steels. Therefore, they cannot rust to give ferrous or ferric oxide, the well-known red/brown rust.

When, however, austenitic stainless steel is heated to above a certain critical temperature, approximately 500°C, in air, the surface can become denuded of chromium, which preferentially oxidizes and thus the critical 18% chrome, 8% nickel analysis no longer exists and the surface can 'rust'. This phenomenon is discussed under corrosion in Chapter 7. It can be identified with the ferroxyl test, which is a sensitive way of identifying free iron – in the form of ferrite.

This test uses a freshly made solution of potassium ferricyanide in acid, which reacts with ferrite to give a brilliant blue colour, Prussian blue. The test is very sensitive and requires some skill and experience to obtain sensible results.

From the above, it should be clear that design engineers and construction personnel should be aware of two important facts. Firstly, when heating for bending, welding or any other reason, there will be the danger of surface contamination in the form of free ferrite. This must be removed prior to the steel entering service. This can be by wire brushing with stainless steel brushes, pickling with nitric, acetic or other acid, or there are proprietary pastes which themselves must be carefully removed after use. This will very often be necessary only in the food industry and certain parts of the chemical industry. There is, however, the danger, which is not great in many circumstances, that crevice corrosion can occur because the surface is prevented from oxidizing.

Secondly, design and quality controllers at the assembly stage must be aware that austenitic stainless steel components should be assembled by careful fitting to reduce tensile stresses. Thus two stainless steel pipe flanges should have their surfaces in compression. That is, they should be pressed together prior to being bolted.

If there is a gap between the flanges and these are brought together by tightening then there will be a tensile force somewhere in the system which could result in stress corrosion. This is discussed and illustrated in Chapter 7.

There are four groups of austenitic stainless steel. To take each of these groups in turn, the failure investigator should be aware of certain characteristics, and whether or not they could have any influence on the problem.

A6a Standard austenitic stainless steels

Specifications BS 970 303/304, BS 970 En58A and AISI 303/304. These can have appreciable amounts of carbons but provided they are annealed and not heated at any stage during manufacture, assembly or in service, and are used under oxidizing conditions, they will give good service. These are basic 18% chrome, 8% nickel, up to 0.2% carbon, and no other critical elements are specified.

These steels can accept appreciable amounts of cold work, increasing the tensile strength, reducing the ductility, and in some instances reducing the corrosion resistance.

When heated in the 300–600°C temperature range for more than a few minutes, chromium carbides will precipitate at the grain boundaries, thus eliminating the single phase structure, and dramatically reducing the corrosion resistance at the grain boundaries – resulting in weld decay. This seriously curbed the use of these materials in their early life. It was then found that by heating to 1050°C and then rapidly cooling, the carbides would remain in solution and thus weld decay was prevented.

This could obviously only apply to components of a reasonable size. It did, however, cure the problem with cutlery knives where the traditional bone handle required the knife blade to have a hot forged tang. Weld decay approximately 10 mm from the tang on the knife blade used to be a common failure.

A6b Weld-stabilized austenitic stainless steels

These are to specification BS 970/321, BS 970 En58E/F/G, and AISI 321. They are either low carbon, less than 0.03–0.05% to eliminate the possibility of chromium carbide forming, or more popularly contain significant amounts of niobium (columbium), titanium or to a lesser extent molybdenum.

The low carbon variety is quite expensive and reduces the tensile properties.

It was found that by adding small amounts of niobium (columbium) or titanium, the problem disappeared. This was because these elements form stable carbides preferentially to chromium which do not migrate to the grain boundary. They thus stabilize the carbides. These steels can therefore be safely heated in the 300–600°C temperature range, thus allowing welding, brazing and forging. The low carbon stabilized steels can present problems when the carbon is near the top limit and the components are held within the critical temperature range for long periods, for example over 0.5 hours.

There are still arguments and opinions on whether niobium is a better stabilizer than titanium, and design engineers and metallurgists have preferences, with the indication that niobium is probably superior to titanium where prolonged heating is involved.

The Strauss (weld decay) test is a positive, simple technique to identify whether or not an austenitic stainless steel is prone to weld decay. With this method, test pieces are sensitized at 650°C for 0.5 hours, then boiled for 72 hours in acidified copper sulphate solution. If at the end of the test the specimen can be bent through 90° without cracking, the steel is 'weld decay stabilized'.

The investigator will readily recognize weld decay with its well-defined grain boundary precipitation, which is generally local and related to welding or local heating. Since heating in the 300–600°C range can result in weld decay, designers must ensure that any process during production or service where a temperature above 300°C is involved must use a weld decay stabilized steel.

If necessary, the degree of the problem can be quantified by producing a number of sensitised specimens for the Strauss (weld decay) test, removing these from the boiling test solution after 24 intervals and bending. If fracture or cracking occurs at 24 hours, the problem is serious. If only slight cracking exists after 72 hours the problem is insignificant.

A6c Molybdenum-bearing austenitic stainless steels

These steels are to BS 970 316, BS 970 En58H, BS 970 En58J or AISI 316. They contain up to 3.0% molybdenum, which stabilizes the chromium oxide and thus makes them more suitable for use in chemically aggressive conditions such as in the food, distilling, brewing, chemical and marine industries. It is particularly important that austenitic stainless steels in these industries are in an oxidizing atmosphere.

The molybdenum stabilizes the chromium oxide and thus increases the corrosion resistance. These steels are now first choice in the food (drink), marine and petrochemical industries.

The molybdenum content has a limited effect on the carbide stability, but not to the same extent as niobium or titanium. Thus steels in this group to be welded, or heated in the critical 300–600°C range, should have a carbon content of below 0.05%, preferably 0.03% max. This is the type 316L – the suffix letter 'L' indicating low carbon.

The investigator identifying crevice corrosion or stress corrosion will require chemical analysis for molybdenum. If this is within specification then the environment, or operating conditions, require investigation.

There is some evidence that the corrosion resistance can be enhanced by heating the surfaces using an oxidizing flame. Crevice corrosion or stress corrosion identified with either covered or contaminated surfaces can be prevented by cleaning the surfaces mechanically then heating with an oxy-acetylene flame.

Where stress corrosion is identified, the stress must be reduced or eliminated before this treatment can be effective.

A6d High temperature austenitic steels

Specifications BS 970 309–310 and AISI 309–310 refer. By increasing the quantities of nickel and chromium up to 20% or more, higher temperature corrosion resistant properties are obtained by the formation of a more stable oxide on the surface.

These steels are used for components in high oxidizing temperatures such as in furnaces up to 400–600°C. They will mostly be used under static conditions, such as in furnace furniture, but may be used on light duty rotating components. Failure can be expected from heat cracking at changes of sections or in thin sections, or in some circumstances from creep. Surface 'rotting' will result from oxidation propagating from the surface.

Relatively small increases above a certain critical temperature can cause failure. The investigator in some instances will be able to report if this is the result of short or long term heating from the extent of the oxidation, and the type of oxide penetration.

A7 Duplex stainless steels

The word 'duplex' means dual or double. The term, when used to describe this group of steels means that two metallurgical phases exist, but does not define with any accuracy the properties which can be expected. The two phases are ferrite and austenite in the majority of cases, but can include exotic inter-metallics.

Investigators should be aware that there is evidence that duplex steels have been oversold. The steels in this group have better tensile strength than austenitic stainless steels, but reduced corrosion resistance.

There are now available duplex steels with specialist analysis, controlled nitrogen content, which require careful heat treatment – solution treatment and ageing (precipitation hardening) – to give mechanical strength and resistance to crevice corrosion and stress corrosion better than existing stainless steels.

The older generation of duplex steels were precipitation-hardening austenitic steels, where the two phases were austenite and a complex precipitate. These particular steels, such as Jessops G18, are no longer produced and will seldom be involved in failure investigation. These were non-magnetic, or only weakly so.

A second series based on 17% chrome, 2–6% nickel, with alloying elements, was produced, typified by the Firth Vickers steel FV520, still in common use. These are similar to modern duplex stainless steels with limited age hardening properties, being truly duplex with austenite and ferrite in variable amounts, generally magnetic.

In the 1980s a new series of duplex stainless steels were produced where the word 'duplex' was used and marketed to a considerable degree. These are based on 17/4 chromium nickel with additives, in particular nitrogen, and other alloying elements, and with chromium up to 25%. They are precipitation-hardening steels and it is claimed they have a considerable advantage regarding stress corrosion, namely, they can be used at higher tensile loads under corrosive conditions, particularly with chlorides present.

These steels are now available as sheets, bars, forgings and castings. There is little doubt that in many cases they are used for components and conditions where they are over-specified, that is, the properties available are not always fully required. This obviously is not good business for the failure investigator as the components will seldom fail in service! There are indications that some of these steels are in service with less than perfect heat treatment, but are not causing problems.

It is claimed that the corrosion properties of these steels are aligned with their mechanical properties, and their mechanical properties are obtained by carefully controlled heat treatment. There is evidence too that steels of this type have had poor heat treatment specified by the user without any history of failure. The author can record that there is some difficulty, even for a qualified metallurgist, in obtaining sensible metallurgical information. There should, therefore, be some sympathy for the intelligent engineer requesting detailed technical information.

In many years of failure investigation the author has never been asked to investigate why something did not fail. There is some justification for efforts being made to investigate why certain steels with the wrong heat treatment do not fail in service as would be carried out in the aircraft engine industry – but not in the oil and petrochemical industry where most of these steels are used. 'If it ain't broke – don't fix it' is the motto and it could be argued that this may be technically satisfactory, but economically it could be quite an expensive policy. An example of this is that there is evidence of steels in service which have had post-weld stress relieving (PWSR/PWHT) for some time at 650°C, which should in theory cause Sigma phase precipitation and resultant brittleness in some, if not all, duplex steels, without serious effect.

Duplex steels are used for shafts, impellers, pumps, casings and closures in the oil and petrochemical industries, with lesser use in general engineering and the food industry.

The problems with heat treatment and the techniques which should be used are discussed in Chapter 14.

The investigator will have identified the mechanism of failure from the original investigation. This could be fatigue or brittle tensile failure which on further investigation is shown to be from corrosion pitting or could be associated with a weld. The welding could thus have contributed to the failure and information on the proposed and actual weld procedure would be investigated. Faulty heat treatment could also be involved, which would require heat treatment trials.

The corrosion potential of these steels relies on the chemical analysis which allows the heat treater to produce the correct metallurgical condition to give the maximum corrosion resistance plus adequate mechanical strength. If corrosion pitting is found therefore, and the investigator is required to identify the reason for it, it will be more economical to carry out a chemical analysis to identify whether or not the material complies with specification, rather than heat treatment and corrosion trials.

It is unfortunate that there can be an insignificant difference in tensile properties between good and bad heat treatment, but a dramatic difference in corrosion resistance. It is time-consuming and expensive to carry out heat treatment trials and corrosion resistance testing. There can also be a dramatic difference in impact properties caused by heat treatment.

The investigator is fortunate that with the expense penalties of these materials, there will almost invariably be detailed documentation regarding the material used, with the chemical analysis and mechanical properties listed, and information on the manufacturing process. Thus it should be possible to evaluate from the visual, macro- and micro-analyses and hardness tests whether the problem is associated with chemical analysis, heat treatment or some other factor such as over-stressing.

There is adequate experimental evidence that a Sigma phase can exist when these materials are held at temperature above 300–400°C for any significant time. This in theory should produce a brittle structure.

It is not unknown for duplex steels to have weld procedures specified calling for pre-heating, controlled heat input and sophisticated prolonged post-weld heat treatment, all of which in theory at least should result in the formation of a Sigma phase. There appears to be no record of failure caused by Sigma phase embrittlement.

B ALUMINIUM AND ITS ALLOYS

Aluminium (Al) is a relatively modern metal not finding significant use until the beginning of the 20th century. It uses the traditional ending of 'um' or 'ium', indicating that it is a metal.

Aluminium in the form of aluminium oxide or bauxite is extremely common on the Earth's surface but is very difficult to obtain by conventional smelting processes. It was not until the discovery of the electrolytic method whereby the bauxite ore is dissolved in molten cryolite (a double fluoride salt) and then electrolysed in the same way as an electroplating process that aluminium metal was produced in quantity. This method results in aluminium being deposited in the molten state at the cathode and copious fumes of fluorides, etc. are given off at the anode. The process requires considerable electrical energy and it was the advent of cheap hydroelectric power which initiated the commercial production of aluminium.

It may be of interest that some of the earliest aluminium was produced on the banks of Loch Ness in Scotland (supervised by the monster?). It is also of interest that the Eros statue in Piccadilly Circus, London, is a very early example of a pure aluminium casting. It is doubtful if any modern foundry would be prepared to manufacture this today, as aluminium is notorious for readily oxidizing and it is extremely difficult to cast pure aluminium. The product of electrolysis is pure aluminium, and this makes aluminium one of the few metallic elements which can be produced from its ore with high purity. Most metals are smelted with considerable difficulty regarding the elimination of the elements found along with the metal.

Aluminium and its alloys are now the second most used metal in the service of mankind, both by volume and by weight. This happened comparatively recently, previously copper and its alloys were the second most common metal. With the increased scarcity and cost of producing copper and the better efficiency of producing aluminium there was a gradual change-over which became dramatic when aluminium was chosen for electricity power conductor wires. Almost all high voltage electrical transmission systems now use aluminium. Although it has a conductivity approximately two-thirds that of copper, it also has a weight approximately two-thirds that of copper, and once the cost of aluminium became less than that of copper there was a dramatic increase in the use of aluminium. Aluminium alloys have also replaced certain grades of iron and steel.

Aluminium and its alloys have good or reasonable corrosion resistance to neutral and slightly acidic environments, and markedly less resistance to alkaline conditions. The corrosion resistance is improved by chromating and anodizing, both of which produce an adherent oxide film which prevents further corrosion provided it is not damaged. More information on corrosion aspects will be found in Chapters 7 and 8. Aluminium is anodic to most metals used by industry and thus must either have an adequate anodic film or paint, or be insulated by a layer of plastic, etc. This applies to all aluminium products. Black, hard, refrigerated or deep anodize are terms which relate to a thicker oxide film, which on pure aluminium is black with a hardness above 500 DPN. With aluminium alloys, the film is grey and less hard.

Aluminium oxide is produced by making the component anodic in dilute sulphuric acid held at 3°C or less. This allows the build-up of a thick 0.2 mm (0.0075″) layer of aluminium oxide. The film is refractory and an electrical insulator with considerable wear resistance when correctly produced. It requires skill and specialist equipment, and the investigator must be aware that the hardness in particular varies significantly between pure aluminium and its various alloys. It finds a variety of uses for wear resistance, electrical insulation and lightweight surface hardened components. The normal anodic film is 0.02 mm (0.00075″) thick, produced at room temperature in the same solution as above.

The oxide is readily damaged. When intact it will act as an efficient insulating layer. It is, however, common that aluminium and its alloys do not corrode under conditions which should cause problems.

Aluminium and its alloys are produced in seven forms. These are:

- high purity aluminium
- aluminium alloyed with manganese

- aluminium alloyed with magnesium
- aluminium alloyed with silicon
- aluminium alloyed with copper
- aluminium alloyed with magnesium–silicon
- aluminium alloyed with zinc

Only the last three are heat-treatable.

B1 Aluminium – high purity

Specifications BS 1470 to 1475 and USA 1000X or 1000 apply with various prefixes.

This metal is produced with 99.0–99.99% purity and is available in wrought form, very seldom as casting. Aluminium has a very strong affinity for oxygen and when heated close to or above its melting point will tend to return to aluminium oxide; hence the difficulties in casting.

The fact that at lower temperatures than 350°C aluminium forms an oxide which is both adherent and coherent adds to its advantage, as the oxide is stable and corrosion resistant.

Pure aluminium is used for a number of purposes, including the modern 'silver paper' and cooking foil. It has low toxicity and can be used in contact with foodstuffs. Aluminium is also used for architectural purposes, sometimes in the high purity form, also as an alloy. It has more or less replaced the 'tin can' for all types of drinks and many food containers.

There is now a suggestion that aluminium contributes to Alzheimer's disease through cooking pots and water treatment. This has still to be confirmed.

A large use of aluminium is for its electrical conductivity, with almost all high voltage electrical transmission using high purity aluminium. It has the disadvantage that it is much more difficult to join than copper as there is no simple soldering process, although modern techniques are becoming available. It is not yet used in household wiring, where brass or copper screw-retaining copper conductors continue to be used, because an aluminium oxide forms which is non-conductive. Soldering, while possible using specialized techniques, is not readily available for the DIY handyman.

Bus bars and similar conductors made of aluminium are now common in industry but at present not for switchgear for the same reason as above, that is, the presence of the non-conductive oxide surface.

There are now available techniques for joining aluminium to copper, thus allowing copper-to-copper joints to be made free of oxidation problems. These include flash butt welding, friction welding and brazing with special fluxes.

It is possible to clad aluminium alloys with pure aluminium, thus combining the strength of the alloy with the corrosion resistance of high purity aluminium.

Aluminium with its oxide is now used as a corrosion protection coating on steel. This can be achieved by metal spray which can then be oxidized by controlled heating. Aluminizing can also be achieved by immersing the steel in molten aluminium under controlled conditions. There is also a sophisticated aluminium diffusion process used on nickel–chrome alloys to improve their hot corrosion resistance.

Aluminium is commonly used as 'anodes' to protect galvanically steel structures such as bridges, ships hulls, etc., from corrosion. The anodes must be in contact with the structure they are protecting. Small additions of mercury appear to improve the anode efficiency.

Failure investigation of high purity aluminium components would be involved with surface contamination, wrong material or misuse. Corrosion of aluminium can occur by damage to the protective oxide by abrasion, contamination or active chemicals.

Chemical analysis might be required to determine the presence of elements such as silicon, manganese, magnesium, iron, etc., where corrosion, electrical conductivity or a ductility problem has been identified. There is no doubt that alkaline marine and industrial atmospheres can contribute to corrosion, but there are many examples of untreated aluminium components used in marine conditions for many years with no corrosion problem. These can be in use alongside identical components destroyed by corrosion. It is more difficult to explain the lack of corrosion than the corrosion!

There is some indication that the initial condition and atmosphere can have a quite dramatic effect on corrosion performance. A warm, dry, neutral atmosphere can produce an adherent oxide capable of resisting a wide range of atmospheric conditions, whereas an initial damp, slightly alkaline atmosphere can result in an active surface which will continue to corrode in service.

B2 Manganese–aluminium alloys

These are manufactured to specifications BS1470–1474/3 or 3000, USA 3000X with various prefixes.

These materials are invariably wrought with the manganese content between 1 and 2%. This results in some reduction in corrosion resistance but permits an appreciable amount of cold working to increase the yield strength and ultimate tensile strength. These alloys are used in large quantities in sheet form.

The London Underground tube trains are almost invariably clad with this and need no painting. Containers used for transporting all manner of goods use this material for cladding, very often with supporting material of other aluminium alloys, although competition exists with austenitic stainless steel, particularly where foodstuffs are involved.

Much of this equipment is now painted as it has been found that spray paint vandalism is easier to remove from the painted surface than from naked aluminium.

Manganese–aluminium alloys are relatively low strength malleable materials which can be used in sheet, tube and bar form for various purposes. Window and door frames, trim on aircraft fittings, household trolleys, etc., are commonly made in this material.

Corrosion resistance can be enhanced by chromate treatment or by anodizing. If it is required to be painted then it must be chromated prior to painting, otherwise adhesion will be very poor.

Failure can be tensile failure from overloading, or faulty or inadequate cold working. Corrosion can be from contamination or galvanic corrosion if in contact with copper, iron (steel) or even zinc (galvanizing). Because of the variable nature of aluminium oxide it is impossible to predict the degree or extent of corrosion.

B3 Magnesium–aluminium alloys

Specifications BS 1470–1474/5 or 5000, USA 5000X with various prefixes apply. These alloys vary from 2 to 10% magnesium.

It is one of the fascinating facts of metallurgy that magnesium and aluminium have a dislike for marine atmospheres regarding corrosion, but when magnesium is added to aluminium, particularly as high as 10%, the resultant alloy has excellent corrosion resistance to marine atmospheres which can be further enhanced by chromating or anodizing. These alloys will also accept more cold work than the manganese alloys and therefore can have their yield strength and ultimate tensile strength improved by an appreciable degree.

Tubular components for golf trolleys, deckchairs, masts, hulls and similar marine equipment would use these alloys as first choice for weight and corrosion resistance.

Any failure in a corrosion context would require chemical analysis for the magnesium only. Experience shows, however, it is more often failure by damage or galvanic effect that results in corrosion. Tensile-type failures would probably require a tensile test if a specification were available.

B4 Silicon–aluminium alloys

Specifications BS 1490–LM2–LMG, BS 1475 N2 apply.

These materials are commonly found in castings. Silicon reduces the melting temperature with a 12% silicon aluminium melting at about 450°C. There are a large number of aluminium castings made from these alloys.

Since the Second World War there has been a dramatic change from cast iron components for items such as switchgear firstly to zinc castings and now to aluminium, with a further trend towards pressed steel components and to a lesser extent plastic taking over from aluminium.

Low strength casings for gearboxes, cylinder blocks and similar items are large users of aluminium–silicon castings. They are also used for architectural and artistic purposes, and there is no doubt that a modern ‘Eros’ would be cast in this alloy and then anodized or chromated.

Failure investigation of these alloys is relatively unusual as they are recognized as low strength, low cost items, with failure occurring as a result of casting porosity, hot and cold shuts, impact or tensile stress. The investigator would use visual, macro- and micro-examinations before considering tensile tests and very seldom chemical analysis. Excessive porosity, shrinkage cracks, 'cold shuts' and other casting defects are not uncommon.

Corrosion failure could be from the environment, or through faulty corrosion protection. Aluminium will 'sacrifice' itself to protect most metals used in industry, thus galvanic corrosion is common.

B5 Copper–aluminium alloys

Specifications BS 1470–1474/11, 15 or 2000x, USA 2000X with various prefixes.

These were the first non-ferrous alloys to be used for high strength purposes. With specific percentages of copper and other elements these alloys can be precipitation hardened (age hardened) at room temperature. This requires that they are solution treated, when they are in the softest possible condition and can be worked, shaped, etc., and then by a relatively low temperature precipitation hardening the tensile strength is increased. These were the original Dural-type materials and this name is still commonly used to cover all age-hardening aluminium alloys. Some of these alloys precipitation harden at room temperature, hence the term 'age-hardening'.

While some casting alloys are available the majority of these alloys are in wrought, bar or tube forms and forgings. These will be found in aircraft, for example in compressor blades and discs, and structural units, and will be the first choice where the heaviest duty, high strength aluminium with reasonable corrosion resistance is required. Transport vehicles which are used round the clock may use these alloys for load bearing components to reduce fuel costs.

'Pop' rivets made in aluminium–copper alloy commonly use the room temperature ageing forms. These can be solution-treated and then refrigerated, thus preventing ageing. After riveting, room ageing will occur, increasing the tensile strength.

Only the aluminium–zinc alloys have a higher yield strength, but they suffer from lower corrosion resistance. Most normal industrial components will specify the magnesium–silicon alloys, which have lower mechanical strength but are cheaper.

Failure can be expected by fatigue from surface damage, sharp radii or corrosion pitting. Damage to any corrosion protection system could initiate fatigue from corrosion damage.

B6 Magnesium–silicon–aluminium alloys

Specifications BS 1470–1474/19 or 20, or 6000x, USA 6000X with various prefixes.

These alloys have approximately 1% magnesium, 1% silicon with or without other alloying elements. They are in effect the 'poor man's Dural', that is, they are age-hardening alloys which do not have as good a mechanical strength as the copper–aluminium alloys but are more economical and find considerably greater use in industry.

These alloys are used on a large scale as the structural members for transport, trucks, trains, etc., where the non-load-bearing components will often be the 2% manganese–aluminium alloys.

Many engineering components, such as connecting rods, extrusions for architectural items, canopies, moving stairs and lifts, where strength plus low weight and corrosion resistance are necessary, use these alloys. The majority are in the wrought form – tube, bar, sheet – with a limited number of castings used for items such as pistons.

These alloys would be the first choice for engineers where weight-saving was important but the components are required to have significant mechanical properties at an economical price.

Failure modes could be corrosion from contamination, chemical attack or galvanic attack. Tensile-type failure would have the mechanism identified by visual examination, aided by macro- and micro-examinations. A mechanical test and chemical analysis might be required to confirm the reason for failure.

B7 Zinc–aluminium alloys

Specifications BS 4300/14 or 17 or 7000x, USA 7000X with various prefixes.

These alloys make up a relatively small group of high tensile alloys which are used where maximum strength and minimum weight are required. They contain up to 7% zinc and the high strength, which can be up to 600 N/mm^2, is obtained at the expense of corrosion resistance. They are used in components such as the landing gear on aircraft and high speed rotating components where centrifugal loading can be high.

These alloys require constant examination for corrosion pitting and fatigue initiation. Corrosion failure would not require further investigation in most instances as the relatively poor corrosion resistance is predictable. The corrosion protection specified would, however, require examination to identify whether this was inadequate or had been damaged.

Tensile failures not explained by porosity would require mechanical testing, with chemical analysis in some instances. Fatigue from corrosion pitting is not unusual.

C COPPER AND ITS ALLOYS

Copper (Cu) is one of the ancient metals discovered before iron, and was responsible for the Bronze Age which followed the Stone Age, preceding the Iron Age.

The Latin word for copper is 'Cuprium', hence the chemical symbol 'Cu'. The reason copper was used prior to iron is that it is easier to win from its ore. Tradition has it that Arabs cooking mutton over a fire discovered bronze when burnt lamb produced carbon, which reduced the copper oxide/copper sulphide/tin oxide 'fire bricks' to bronze.

Copper is still used but is being replaced by zinc plated steel, austenitic stainless steel, to some extent plastics/polymers, and also aluminium. Aluminium can be anodized to give a very good imitation of copper and many of the copper alloys, depending on the dye used in the anodizing process.

Such is the advance of metallurgy that it now requires the metallurgist to be capable of identifying whether the duchess is wearing a solid gold, copper alloy, aluminium colour-anodized, or gold plated plastic tiara!

Many of the copper alloys were produced and developed because of the armaments industry and the names reflect this unfortunate episode – cap copper, gunmetal, cartridge brass, naval brass, etc. There are fewer generic specifications for copper and its alloys than for other metals, thus there is no attempt to supply this information. The metal and its alloys are still referred to by the traditional names.

There is an increasing shortage of copper ores, and the higher cost of winning the metal from the ore, compared with aluminium which is more plentiful on the Earth's surface, has spelt the decline of copper and its alloys. There still remains an important place for copper, however, and copper at present is the third most common metal used in industry either by volume or by weight. There is little doubt, however, that if the price of titanium was significantly reduced then copper would move to fourth place.

Copper and its alloys still find use where electrical conductivity is involved, and easy joining is essential. While aluminium is competing successfully in the industrial field it is not yet possible to join aluminium in a satisfactory long term condition as simply as copper. Aluminium joints when warm form a non-conductive oxide which builds up until there is a low conductivity at the joint, with serious fire risk due to localized overheating. With copper and brass the DIY electricians can produce joints in copper which could not be achieved with aluminium.

Failure of copper and its alloys could be expected to follow the same pattern as other metals. That is, where corrosion occurrs it will be in the form of surface discoloration with what used to be known as 'season cracking'. The modern metallurgist would call this stress-corrosion cracking and all copper alloys, with the possible exception of copper beryllium, are susceptible to this failure mechanism.

Brasses are subject to de-zincification, which is where the zinc is preferentially corroded. This is visually unsightly and can also result in stress raisers at grain boundaries which will encourage fatigue propagation.

All copper alloy castings can have porosity, and aluminium bronzes are particularly difficult to cast because of their affinity for oxygen.

Fatigue, tensile stress and torque/shear characteristics can be seen as readily with copper and its alloys as with other metals.

Galvanic corrosion is unlikely with copper and its alloys as copper is high in the electrochemical series and thus all common engineering materials will sacrifice themselves to protect it. Thus any corrosion problem identified as possibly galvanic with a copper or copper alloy would point the investigator to electrolysis where some stray direct electrical current converted the copper component into an anode in the presence of an electrolyte. This appears to be a more common problem than is often appreciated.

C1 99% Copper

This is produced in several forms.

C1a Electrolytically-refined oxygen-free copper

This is copper which has been smelted to the 'blister' stage and then electrolysed under controlled conditions to give 99.9% copper with no oxygen. This has an electrical conductivity of 100% IACS – International Annealed Copper Standard. Only pure silver has slightly better electrical conductivity than this version of copper.

This is the material which is almost invariably used for copper electrical conductors, particularly in high voltage electrical transmission and the electronics industry, but also in normal electrical wiring contracts. Failure can be expected because of its low strength. It should always be used in the annealed condition. Any cold work will reduce the electrical conductivity.

This material is produced in the form of wire, bars and strips. It would never be used as a casting except under very exceptional conditions which would require vacuum casting as it has a very high affinity for oxygen. It is not possible to weld this brand of copper as it will produce copper oxide even using techniques such as inert gas welding. Any quality welding would require to be carried out under high vacuum.

C1b Phosphorus/arsenical/fire-refined copper

These forms of copper, which are 99% copper or higher, are produced from the initial smelting stage which is normally called 'blister copper', and have certain quantities of oxide, sulphide and other impurities with no useful properties. Any one of the three techniques using phosphorus, arsenic or fire refining can produce high purity copper which will contain minute amounts of oxide, which reduce the electrical conductivity.

There is no visible difference, however, between these coppers and the copper produced by electrolysis. Thus for visual purposes such as decoration, and use in industry where the corrosion resistant properties (or any other property apart from electrical conductivity) are required, these are useful materials and are cheaper to produce than electrolytic copper. These materials used to be commonly used in the food industry, and are still used for stills where whisky is involved. (These are the vessels which make the famous 'Scotch' whisky.) It may be of interest that there are minute quantities of copper in Scotch whisky and tradition states that this is necessary to give the unique taste to the whisky along with the water and air of Scotland. This means that the stills do not have an infinite life and a significant industry exists in Scotland manufacturing copper stills.

These coppers are still used for piping in plumbing but are being replaced with austenitic stainless steel and plastics/polymers. One rather sad comment is that the change from copper to austenitic stainless steel was accelerated by the fact that thieves did not recognize that austenitic stainless steel is more valuable than copper. Thus copper had a higher security risk, which increased its cost over austenitic stainless steel on building sites, etc.

As implied in the preamble, the use of copper in jewellery, while it still exists, has largely been replaced by other metals such as anodized aluminium and now anodized titanium, along with the electrodeposition of very thin gold deposits as a quality finish.

C1c Low alloy coppers – up to 1% alloy content

There is an alloy containing approximately 1% cadmium which enhances the ability of copper to accept cold work without markedly reducing the electrical conductivity. This 1% cadmium copper is therefore used for overhead electrical and telephone cables. This allows longer distance between the pylons.

Where tensile failure has to be investigated, analysis and mechanical tests might be necessary.

Copper alloys containing small quantities of silver, zirconium, tellurium, boron and manganese are produced for various specific properties. Any failure of these specialist materials would require analysis in most instances once the mechanism of failure had been found, and the reason for failure not identified.

C2 Brass

Brass is an alloy of zinc and copper. There are two distinct alloys, alpha and beta brass.

Brasses were the materials used in the Industrial Revolution for vast amounts of fixtures, fittings, nuts, bolts, bars, etc. Most have now been replaced with electroplated mild steel, austenitic stainless steel, 12% chrome stainless steel, etc.

At one time brasses may have been the most common metal used in industry. They have excellent corrosion resistance in normal environments, reasonable mechanical properties, particularly ductility and impact resistance, and are readily machined.

C2a 40/60 alpha brass

This contains 35–40% zinc, 60% copper and is the brass for nuts and bolts used at room temperature, produced from bar, very often cold worked to give a round, hexagonal or square shape, and also brass tube.

Alpha brass accepts considerable cold work. The material known as 'cartridge brass' could be cold drawn to produce a cartridge for shell cases with a thin wall and considerable depth, up to 1 metre in some cases. This was done without inter-stage annealing.

Many drawn components, such as hollow vessels, were therefore manufactured in this material. It is still found on a limited scale. Failure could be de-zincification or thinning at the drawing operation. The metallurgist, without either chemical analysis or mechanical tests, should be able to identify the mechanism of failure.

It may be necessary at times to carry out analysis for zinc to identify that the correct amount is present as this could contribute to failure. These materials cannot be hot worked without becoming brittle. While there may be some evidence of this, the failure investigator would probably require analysis to confirm it.

De-zincification is a common problem where the zinc is denuded, resulting in a surface weakness which can propagate either by stress corrosion (commonly known as season cracking) or by fatigue. This can be recognized by the yellow colour of brass disappearing either locally or all over, and the red colour of copper appearing.

This group of alloys would include 'gilding metal', which has approximately 15–20% zinc, a gold appearance and low electrical conductivity.

With the addition of approximately 1% aluminium, 'naval brass/admiralty brass' is produced. This has the different colour of yellow to standard brass, and has good corrosion resistance to sea water. It was commonly used in ship condenser tubes although this has largely been replaced with austenitic stainless steel.

An investigator identifying corrosion would require an aluminium analysis. This should be at a level of approximately 1%.

C2b 30/70 Copper–zinc beta brass

This has the opposite characteristics of alpha brass when it comes to working. This material cannot accept any significant amount of cold work, becoming brittle. It can, however, be readily forged and hot formed.

While there remains considerable quantities of 70/30 brass components in service, it is unlikely that a modern designer would specify this material. Replacements in aluminium, austenitic stainless steel, 12% stainless steel or mild steel electroplated would be the economical choice.

Failure investigation of beta brass components would identify whether or not the material had been subjected to cold work which could then accelerate failure by fatigue.

Like alpha brass, it would probably be necessary to carry out an analysis for zinc to satisfy the failure investigator that he/she is on the right lines.

Both alpha and beta brass can have free machining constituents added, the most common being lead. These result in excellent machinability with very limited tool wear and there are records of batches of brass bars being set up in semi-automatic or automatic machines which ran day and night for many years without a tool change, producing components such as nuts and bolts.

The term 'manganese bronze' applies to alpha and beta brass. Approximately 1–2% manganese is added and gives an attractive bronze colour, with some increase in tensile strength. The use of the term 'manganese bronze' would appear to have initiated the idea that any copper alloy not a brass was a bronze.

The names manganese, beryllium and chrome bronze were at one time common. It is now accepted generally that the word 'bronze' is related to the tin–copper alloy, and others are copper alloys, for example beryllium copper.

C3 Bronze

Bronze traditionally is a tin–copper alloy. At one time there were three forms of copper and its alloys – copper, brass and bronze.

When it was discovered that by adding some manganese to brass, the strength and corrosion resistance could be increased and the colour became slightly more brownish than brass, then the alloy was given the name 'manganese bronze'.

Tin-based bronze, the subject of this group, contains from under 1% to over 10% tin. It has an attractive brownish colour and excellent corrosion resistance, particularly to marine atmospheres, which is proportional to the tin content. It contains other alloying elements such as manganese and the castings traditionally include zinc, lead or phosphorus.

The most popular casting alloy of all time is probably the 5/5/5 tin–zinc–lead–copper known as 'gunmetal'. This has excellent fluidity, pours without problems and is not susceptible to shrinkage or oxide porosity. While still used on some occasions, fabrication in electroplated steel and castings in aluminium, zinc and grey cast iron are all economical alternatives if sufficient quantites are required. Plastics/polymers are now also being chosen in similar applications to those for bronze.

Tin bronzes are still used for bearings, particularly heavy duty low speed bearings, where good marine corrosion resistance and attractive appearance are required. These now have competition from PTFE, sometimes impregnated with bronze.

Failure investigation of a bronze would identify the mechanism of failure, for example in a bearing evidence of heat could indicate lack of lubrication, evidence of tearing could indicate either break-up of the bearing or excessive cold work. Galling can occur where heavy load, poor lubrication or some other problem results in a cold weld of the contact surfaces.

Visual and micro-examinations can identify evidence of surface heating, abrasion, cold work, or fracture of the galling weld.

Depending on the mechanism, the failure investigator might require analysis for tin, lead or zinc. The proportion of these metals gives an indication of the standard of bearing material involved. In general the higher the tin, the lower the lead and zinc, then the better quality the component, particularly regarding bearing properties and corrosion resistance.

A limited amount of bronze plating is carried out where interference fits are required. This at one time was quite popular in the aircraft engine industry, where it was discovered that a deposit of bronze on taper bolts ensured an excellent fit. This would only operate successfully where no electrolysis could occur resulting in the tin bronze causing corrosion of steel components. Both tin and copper will accentuate corrosion of steel when a conductive fluid such as water is present. This might be a factor in any investigation where tin bronze plating had been used and steel corrosion occurred.

C4 Cupro-nickel: nickel–copper alloys

This material must not be confused with Monel. Monel is a nickel alloy containing copper. The copper–nickel/cupro-nickels are copper alloys containing nickel.

Nickel–copper alloys contain from 1% to approximately 10% nickel. They have a copper colour, excellent deep drawing and marine corrosion resistance. They are quite expensive and while their use is increasing to some extent in the offshore oil industry, they are a specialist application and will seldom be involved in failure investigation.

C5 Nickel-silver: nickel–zinc–copper alloys

These are alloys of copper with nickel and zinc, the nickel and zinc totalling approximately 25%, with higher nickel content improving the appearance and corrosion resistance.

These alloys will seldom be found in modern industry as they have been almost eliminated by the use of austenitic and 12% chrome stainless steels. They were used on a very large scale in the cutlery industry where the production of a white metal, that is nickel silver, with reasonable corrosion resistance could be used for the manufacture of cutlery. This cutlery could then be electroplated using silver, or nickel chromium, and when the surface coating was removed by abrasive polishing, a white metal appeared as distinct from brass. EPNS stands for ‘electroplated nickel silver’.

Nickel silver is still used to some extent for jewellery or ornamental purposes and a very limited amount is used for specialist purposes in industry where, for example, electrical conductivity and corrosion resistance might be required. The conductivity varies from 10 to 15% of pure copper but the contact resistance remains constant.

The investigator would almost invariably require analysis for zinc and nickel to identify if they were in the correct ratio. This is not a material which could be expected to be used or recommended by the modern metallurgist or engineer, and will seldom be involved in any failure investigation.

C6 Aluminium bronze: aluminium–iron–nickel–copper alloys

These alloys, although invariably termed 'bronze', contain no tin. They will almost invariably contain approximately 10% aluminium and 5% iron, with the better alloys containing up to 5% nickel.

These materials have excellent corrosion resistance, particularly in marine atmospheres, with good mechanical strength and excellent abrasion resistance. The aluminium results in the formation of aluminium oxide (Al_2O_3) at the surface. This is extremely abrasive, being the grit used for blasting and high quality grinding wheels. There is on record a problem where an aluminium bronze gear running with a nitrided steel gear in a fuel pump rapidly wore through the nitrided case.

Aluminium bronzes are available as bars, forgings and castings. It is notoriously difficult to cast, being very prone to 'hot shuts' where the aluminium oxide forms on the molten surface of any splash or run of metal, preventing intermingling. This can be a cause of failure, readily identified in a macro- or micro-specimen.

These materials are in the fore of copper alloys – and still the first choice in marine atmospheres, where long life, low maintenance and high integrity are required.

Failure could be by tensile failure, impact, and in some circumstances corrosion, but this would be unlikely. Chemical analysis would supply useful information provided a technical specification was available. The presence of nickel will always ensure improved properties, thus its absence could contribute to failure.

C7 Beryllium–copper (beryllium bronze) alloys

Until recently this material was known as beryllium bronze, but the modern tendency is to refer to it as beryllium copper.

These alloys have the highest tensile properties in the non-ferrous range. They can have strengths up to 1500 N/mm^2 (100 $tonf/in^2$).

There is a narrow range of chemical compositions covering all of the beryllium coppers. This is typically 1.5% beryllium and up to 2% cobalt, the remainder being copper. They are precipitation hardening (age hardening) alloys, solution treated at approximately 750°C when they are in the softest possible condition, and can be readily worked. They have an electrical conductivity of approximately 20% IACS in this condition.

When they are precipitation treated (aged) at a temperature of approximately 450° C they will be in the fully hardened condition with the inter-metallic compounds precipitating to give low ductility and high tensile strength. This is the condition which is required where tensile properties alone are specified. These alloys are used for components such as hammers, spanners, etc. used in explosive or flammable conditions. Being non-ferrous there is no danger of sparks when impacting steel with a beryllium copper hammer or spanner and thus petrol, gas, ammunition and explosives stores and manufacturing procedures make use of tools in beryllium copper.

They were used at one time to a considerable extent as draught excluders on doors where the high tensile, non-corrosive spring gave long life. This has now been replaced by 12% chrome stainless steel which is much cheaper. The more widespread use of central heating has also reduced the need for draught proofing.

A considerable amount of beryllium copper is, however, used in the electronics/electrical industry, where a conductor and spring are combined in the form of switches and contacts. These components are manufactured normally from strip material, are shaped in the solution treated condition, and are then over-aged, that is, they are taken to a temperature slightly above the ideal precipitation hardening temperature, or held for a longer time than would be ideal. This reduces the tensile strength of the component but increases its electrical conductivity. Thus there is a compromise regarding tensile strength, which is adequate for mechanical requirements, and improved electrical conductivity. The conductivity is increased from approximately 25% IACS at maximum tensile strength to 45% at 620 N/mm^2 (45 $tonf/in^2$).

The failure investigator with a fractured component would require to identify firstly the mechanism of failure, then the hardness, and would then perhaps require analysis for beryllium if there were any doubt regarding the tensile properties. The alternative would be to carry out heat treatment tests on the failed component if the tensile strength was below that specified. It is claimed that this material is not prone to stress corrosion.

There are now examples of beryllium copper being used in industries such as the oil industry, where non-spark, corrosion resistance and high strength characteristics justify the cost of this relatively expensive material.

There are components where the critical piece is manufactured in beryllium copper which is then welded using special techniques to 12% chromium steel, duplex steel, or in some instances normal alloy steels. This requires detailed specification information and good metallurgical control.

Beryllium itself has a serious health hazard but there is no indication that the use of beryllium copper presents any health risk.

C8 Chromium–copper (also known as chromium bronze) alloys

There are a small number of alloys with up to 1% chromium, which increases the tensile strength with a minimal reduction in electrical conductivity. This is not as dramatic as that found with beryllium copper, but components such as electrodes for welding make good use of these alloys.

These are age-hardening alloys which retain their hardness up to 350°C. They are used for clamps at flash butt welding, electrodes for spot and seam welding, etc. The chromium, in addition to giving hardness, also resists oxidation, thus any arcing between the electrode and the component has a minimal effect on the contact resistance.

It is unlikely that many of these components would ever be submitted for failure investigation. Any investigation would involve identification of the mechanism of failure – this would most likely be poor conductivity or low tensile strength, in which case chemical analysis for chromium would most probably be required.

C9 Copper alloy braze materials

Until the quite recent past brazing was a very common method of joining metals. A definition of brazing is that the metals being joined do not melt but the metal used as the joining material melts and forms an inter-metallic bond with both the components. It is thus a form of glue and is therefore quite different from welding where there is a melting and mingling at the join of all components involved.

Soldering and brazing are technically similar. The term 'braze weld' should be treated with caution as it generally means brazing but on occasion can be correct in that a copper alloy may be welded to a copper alloy. Even so, there does not seem to be a valid or accurate definition of the term 'braze weld'.

Brazing requires that the material used for joining melts at a lower temperature than the components being joined. Until recently braze metals were almost all alloys of copper. These included the silver solders for which alloying with silver and cadmium reduced the melting point of the joining metal to as low as 400–450°C. Traditionally the braze metal melted in the region of 850–950°C, which is below the melting point of steel. Braze metals have considerable variation; many of them are forms of brass, some are more like bronzes, and there are complex alloys with silver and cadmium. The high cadmium silver solder braze metals are used much less at present as it has been found they present a health hazard.

The word 'braze' does not necessarily mean copper alloy and there are now available braze materials based on nickel and some with the precious metals gold, platinum and palladium.

The important aspect with a braze metal normally is its melting point and the fact that it has a range of temperatures over which it solidifies. It takes less skill to produce a good braze where there is a range of temperatures over which solidification takes place than where there is an instant transformation from liquid to solid.

The ideal condition for brazing is that there should be a very tight tolerance regarding the sizes of the two components being brazed. If it is a male/female type braze then the clearance for the best quality braze is in the region of 0.02 mm. This means that the components for high quality brazing must be machined to a much tighter tolerance than is normally found in engineering.

With lap-type joints the two surfaces must have a high quality finish in the region of 10 micro-inches or better for good brazing. If the surfaces are placed in contact without being clamped then ideal conditions will exist.

Brazing ideally should be carried out with the metal fed from the bottom. The component should be preheated to somewhere near the brazing temperature and then the braze with a flux fed from the bottom. With ideal conditions a braze metal will rise by capillary action as much as 100 mm.

With the advent of modern welding very few skilled brazers remain. Brazing, however, still occurs in high productivity component assembly using either vacuum brazing or brazing in a controlled atmosphere such as hydrogen in a brazing furnace. In this instance the components are assembled, very often with the braze metal being applied in the form of a paint or paste where the braze metal is in powder form together with the flux. These parts are passed through the furnace on a conveyor at the required temperature. Vacuum, hydrogen, hydrocarbon and nitrogen are all used as atmospheres for brazing.

The visual examination would show if the failure was of the braze metal, the braze interface, or remote from the braze. Micro-examination would then indicate if the braze was of the necessary standard. It may be necessary to carry out chemical analysis of the braze metal if problems are identified regarding lack of fusion.

D ZINC AND ITS ALLOYS

Zinc ore exists in pockets in the Earth's surface and the metal is obtained by conventional smelting techniques which can present considerable health problems. It is available in a reasonably high purity form and can be refined where high purity zinc is required, which is rare, apart from the anodes used for electroplating.

Zinc (Zn) is used as anodes for the cathodic protection of steelwork, and there are alloys of zinc which are designed more for castability than for their mechanical properties, although there are a few alloys where mechanical properties are enhanced.

D1 Zinc – commercially pure (99%)

The use of commercially pure zinc is confined to objects such as anodes for protecting steelwork. Zinc and aluminium are the common materials used, with aluminium being theoretically more efficient as an anode than zinc, but preference is given to zinc in many instances because of possible problems with aluminium.

The only reason for failure investigation of zinc anodes would be that steel corrosion was occurring. This would almost invariably be found to be the result of faulty connection between the zinc anode and the steel cathode.

Unlike aluminium, zinc does not produce a self-healing non-conductive oxide, and it is unlikely that the anode would fail becuase of poor conduction. If the anode is found to be corroding too rapidly, this might be because an alloy has been used in place of pure zinc and this would require chemical analysis. The investigator might also consider, and look for, the presence of stray electric currents causing electrolysis.

Zinc is probably the most common metal used as a surface coating to protect steel. It is applied as an electrodeposit with a thickness of 0.01 mm maximum. Most heavy duty components now require that the electrodeposit is 'chromate passivated' since there is evidence that this improves the corrosion resistance by at least 100%. This applies to the yellow, dull chromate finish where the visual appearance is variable, and can be unattractive. The blue/clear chromate will enhance the corrosion resistance, but not to the same extent. The electroplated deposit is sometimes referred to as 'electrogalvanizing'.

Galvanizing is the hot dip process, where the cleaned etched steel component is immersed in molten zinc. This should have a deposit of at least 0.1 mm, which is ten times that of the electrodeposit. Like the electrodeposit the galvanizing can have the corrosion protection improved by chromate passivation. With a small addition of aluminium to the zinc, the resultant deposit will be bright.

Hot dip galvanizing deposits have considerable variation in deposit thickness, and are not normally acceptable for fine limit components. Nuts and bolts are commonly machined after galvanizing, which can result in removal of the protective coating.

Sheradizing is a 'pack galvanizing' technique where components are heated along with zinc granules in a controlled atmosphere. This produces an adherent zinc film up to 0.1 mm thick which follows the contours, giving a homogeneous deposit unlike the hot dip galvanized product. Zinc is also deposited by metal spraying.

A failure investigation would readily identify the type of surface deposit and its thickness by micro-examination. Chemical checks would find if chromate passivation had existed.

It is essential that any zinc deposit which has to be painted either has some form of pre-treatment or has an active primer applied. Thus an investigation into a paint failure on a zinc or zinc coated component would require further tests. If the paint film can be removed in large flakes, it can be stated that the preparation was faulty.

Zinc components being relatively cheap are seldom involved in failure investigations. The components can be expected to have low mechanical strength, low ductility, and poor impact strength. Thus failure of components is to be expected in the hands of domestic users even with normal handling. Failure investigation would identify the mechanism of failure, and would then look to see if the casting was of a satisfactory standard. Any excessive porosity or oxide could mean that this material was below strength, but there would seldom be any justification for carrying out chemical analysis, or mechanical tests.

D2 Zinc alloys

The vast majority of zinc alloys are used as castings with die castings being the most popular type. Zinc alloys are probably the cheapest materials for producing an accurate casting but they have limited strength and low ductility.

Zinc alloy castings are therefore are used for products such as toys, low strength household components, light fittings, vacuum cleaners, camera bodies, tape recorders, and auxilliary fittings on cars, bicycles, etc.

Zinc alloys have replaced aluminium in certain areas in which aluminium had replaced cast iron. Zinc die castings are now finding competition from modern plastics/polymers and also where high production is possible from pressed sheet metal components.

Many zinc alloys are electroplated to simulate other products which were commonly chromium plated. It is more difficult to electroplate zinc than copper alloys or steel, but easier than plating aluminium and its alloys. Any coating porosity, however, allows the atmosphere to attack the base zinc, forming the oxide which being bulky produces unsightly blisters.

The zinc alloys require the same special preparation prior to paintaing as zinc itself.

E NICKEL AND ITS ALLOYS

Nickel is a comparatively modern metal and is relatively rare on the Earth's surface but is found in high concentrations in specific locations. The chemical symbol is Ni.

In North America and Australia, nickel deposits are found in conjunction with copper and when some ores are smelted the 60/40 nickel–copper alloy known as Monel is produced. This was the first nickel alloy used in industry.

E1 High purity nickel – 99%

High purity nickel can be obtained by the Mond process whereby 'nickel carbonyl', which is a gas produced from the unpurified nickel by combination with carbon monoxide, is passed up a column where nickel is 'seeded out' as high purity metal. This is a rare gas-to-solid process in metal production. High purity nickel is also produced using electrolysis.

Until the advent of low cost austenitic stainless steel, nickel was commonly used as an electrodeposited coating on other materials such as mild and alloy steels, copper alloys, etc., to give excellent corrosion resistance. Many of the advances in the petrochemical industry were achieved by the use of these electroplated components. Very little is now used as it has been replaced by austenitic and 12% chrome stainless steels and, to a lesser extent, titanium.

High purity nickel, therefore, is seldom found in industrial applications. It is still used but on a reducing scale in the electroplating industry, where it is the essential underlay in decorative chrome plating.

Alloys such as cupro-nickel are discussed under the copper alloys, and nickel is a common constituent in alloy steels and also is an essential constituent in austenitic stainless steels.

Nickel is a comparatively expensive element and thus its use must be justified. The most common nickel alloys in use at present are the 80/20 nickel chromes where the creep resistance, that is, the high temperature strength over a period of time, is excellent. They are used for components in gas turbines and other high temperature equipment where the higher the temperature, the higher the efficiency of the equipment. The 80/20 nickel–chrome alloys also have excellent heat oxidation resistance.

The 60/40 nickel–copper Monel alloys are now comparatively rare as a first choice, having been replaced by austenitic, martensitic or duplex stainless steels, or where higher temperatures are involved, nickel–chrome alloys. They still find some uses in aggressive marine conditions, where corrosion damage can be very expensive.

The failure investigator would identify the failure mechanism, and where tensile failure had been identified would look for ductility, brittleness, creep, hot rupture or fatigue. Once the mechanism has been identified a report will be issued or further work carried out. This might require scanning electron microscope (SEM) work to identify the corrosion product at the surface or grain boundaries.

Pure nickel will seldom be found with modern equipment. As an electrodeposit it found considerable use in the chemical and food processing industries, but has been largely replaced by type 316 austenitic stainless steel, nickel alloy, or titanium alloy, and in some cases plastics/polymers.

Electroless nickel plating, which is a surface deposit of an alloy of nickel and phosphorus, is quite common in the petrochemical industry as a corrosion protection, and also for abrasion resistance. The deposit can be age hardened to give a surface hardness in excess of 400 DPN. This was originally known as 'Kanigen plating' but there are now various other solutions used.

The failure investigator would require to identify the failure mechanism, adhesion (tensile), wear (abrasion), or corrosion. The deposit can be very thin, thus presenting problems for hardness testing. The scanning electron microscope (SEM) might be required to confirm the findings of the optical microscope.

E2 20% chromium–nickel alloys – non-age hardening

Non-ageing alloys were commonly used for electrical resistance heaters and their ability to operate in hot oxidizing conditions. Rarely used today. Based on 80/20 nickel–chrome.

E3 20% chromium–nickel alloys – age hardening

These, like the alloys in the previous section, are based on 80/20 nickel–chrome but have a number of other elements added which form inter-metallics which can be taken into solution and then precipitated. More information is given on this in Chapter 14 on Heat treatment.

The elements involved include aluminium, cobalt, titanium, iron and molybdenum. In exactly the same manner as for other age hardening alloys there is a solution treatment, which is performed between 1050 and 1200°C, followed by quenching and then a long term precipitation treatment, also known as 'ageing'. This will be in the region of 700–850°C.

While this will increase the tensile strength of the alloys to a limited extent, the gain is not normally as dramatic as with other age-hardening alloys. It does, however, improve the creep strength. Creep is where a component held at temperature with or without a tensile load will stretch after a long time, although the load applied is below the proof/yield strength of the material.

It should be obvious that with gas turbines or jet engines, the energy obtained is from the hot, high velocity gases which impinge on the turbine blades. These operate well above red heat and at high speed, thus with significant centrifugal load. The nickel–chrome precipitation-hardening alloys used have unique hot tensile/creep and corrosion resistance.

It was the advent of these metals which allowed Whittle to successfully develop the first gas turbine engine. It is recorded that Leonardo de Vinci in the 15th century proposed the theory of jet engines but it was not until high creep resistance materials became available that the idea could be exploited.

The failure investigator involved with these components will have identified whether the problem is tensile stress, corrosion or over-heating.

While nickel–chrome alloys have excellent resistance to oxidation at high temperatures, their resistance to sulphur contamination (particularly if this is involved with chlorides and moisture) is poor. This also applies to vanadium contamination. Thus a failure investigator with a corrosion problem might require the services of the scanning electron microscope to identify the contamination at the surface or grain boundaries.

Where tensile failures are involved, the failure investigator will look first for evidence of creep. This requires specialist knowledge and etching techniques where the creep voids which occur at grain boundaries can be positively identified. It is of vital importance to differentiate between creep and hot rupture, which is where an excessive temperature has been involved, or where the tensile load has increased for the same temperature. This will be a different metallurgical effect than that of creep, which is a long term defect. This will also differ from straightforward tensile failure or fatigue. That is, if it is hot rupture this could be some instantaneous effect which occurred, whereas creep is a metallurgical effect which takes many hours at temperature. It will thus be appreciated that from the client's point of view it is essential that the reason for failure is identified and reported. This can only be found at magnifications up to x500.

While these alloys are in the main involved with gas turbines, there are other engineering products which make use of their high temperature/high oxidation resistance. These might not be rotating components but use their tensile properties at elevated temperatures.

E4 Copper–nickel alloys (Monel)

These alloys were identified first as Monel. There are nickel–copper ores, particularly in Canada, which when smelted result in a 60/40 nickel–copper alloy. Known as Monel, it was the first high temperature alloy available to engineers.

Monel is now being superseded in many instances by austenitic stainless steel, 12% chromium stainless steels, and to a lesser extent titanium alloys, the 80/20 nickel–chrome alloys, and duplex steels.

Monel has excellent corrosion resistance, particularly in marine/sea water conditions even at high temperature. It is, however, expensive and while still used to some extent in sheet form and for certain pump parts and other mechanical pieces, the use of Monel is much smaller now than in previous years, although there is a tendency for the offshore oil industry to specify Monel for shafts, in pumps for aggressive conditions.

Monel failure would be expected to be either through corrosion or tensile stress. If from corrosion then the investigator would look for high temperature effects and discuss the atmosphere in which the material had been operating. The possibility of stray electric current causing electrolytic corrosion must always be considered.

There could be occasions where the use of the scanning electron microscope (SEM) would be desirable to examine the surface or the grain boundaries, to identify a corrosion product. Like other nickel alloys sulphur compounds can cause problems. Stress-corrosion cracking, while rare, has been identified with these alloys.

Where tensile failure is involved, the mechanism is important. Fatigue propagation would require the investigator to identify the reason for initiation. This might be surface damage or a sharp radius, but could also be grain boundary attack or stress–corrosion cracking. Creep or hot rupture would also have to be considered.

The investigator would want to identify the type of Monel involved, and whether or not this was of the age hardening type. There are a limited number of alloys in this section which can be solution-treated and aged. This, as with the nickel–chrome alloys, does not make a dramatic difference to the tensile strength, but the difference might be important in a failure.

E5 Iron–nickel alloys

These make up a range of materials, many of them in the 50/50 nickel–iron range. They are used because of specific magnetic properties and also their thermal expansion characteristics.

By certain types of very specialized heat treatment, certain magnetic properties can be produced but these would not often come within the orbit of the normal failure investigator. Specialist advice and information regarding the type of heat treatment would be required.

A limited number of versions are used for their good hot oxidation resistance.

Mechanical failure would require the same skills as for any other metal.

Some alloys are produced for their specific expansion properties. There is a series which has the same expansion characteristics as glass and thus can be used with glass/metal seals. Secondly, there are alloys which have a very low coefficient of expansion, used in components such as bimetal switches.

As with the magnetic alloys, specialist advice would be necessary where a failure occurred through expansion/contraction problems. The investigator may require to carry out controlled heating using a dilatometer to graph the expansion/contraction against temperature of the parts involved to check for compatability.

Mechanical failure, which could be expected to be unusual, would require the same skills as for other mechanical failures.

E6 Beryllium–nickel alloys

There are a small number of alloys with low quantities of beryllium which are analogous to the beryllium copper alloys. These have much lower electrical conductivity but better temperature properties. They also find uses in bimetal switches and other control mechanisms where a stable oxidation-resistant material with hot strength is required.

They are not commonly found in industry, and can present problems to the investigator in that the difference between good and bad heat treatment (solution treatment and ageing) can be slight, while the variations in operating conditions can be considerable. The investigator thus has the quandary that a minor deviation from specification in mechanical strength at temperature might be blamed for failure, whereas in fact the component had been slightly abused in service.

Most components in this group will be very small mechanical parts, subjected to light loading, but expected to operate sophisticated equipment.

F TITANIUM AND ITS ALLOYS

Titanium is a 'modern' metal. It is also one of the most common elements on the Earth's surface but in the form of titanium oxide, which is difficult to win from the ore. The chemical symbol is 'Ti'.

Titanium oxide (titania, TiO_2) is a common pigment being a very stable white. It is used in place of zinc oxide and other white metallic oxides which tend to discolour, particularly with sulphides. It is also inert with high temperature properties and therefore is preferred to organic pigments.

The extraction of titanium from the ore is extremely difficult because it requires complex high temperature processing using chlorine and other gases to form a titanium 'sponge' which is the initial impure titanium metal. This must then be melted under high vacuum to produce titanium metal. Forging or hot working of heavy duty components is carried out with the surfaces protected by molten glass.

At room temperature and up to 250–350°C titanium has excellent corrosion resistance against all known chemicals. It is the wet chemist's nightmare in that it is very difficult to dissolve using standard acids. Titanium therefore finds use where aggressive chemicals are used in the petrochemical and similar industries at temperatures not in excess of 300°C. Gas turbine engines can have several stages of compressor blades in titanium alloy.

Unfortunately, depending on several factors, with temperatures above 250°C, titanium will react with many elements, forming oxides, hydrides, nitrides and carbides. These form at the surface and diffuse into the metal, considerably reducing its corrosion resistance and mechanical strength. The metal therefore is of limited use above 300°C. Reactions are time/temperature dependent, that is, a longer time at a lower temperature will give the same reaction as a shorter time at a high temperature.

Titanium can be anodized to give improved corrosion resistance. This increases the temperature at which it can be used by a limited amount, but this improvement is lost if the anodic film is damaged, and this can result in a stress raiser to initiate fatigue cracking.

By altering the surface finish, and anodizing conditions, various surface colours can be produced.

Titanium and its alloys are found in three separate forms.

F1 99% titanium

Commercially pure and high purity titanium is available in bar and sheet form and is responsible for the highest usage of the metal.

Flame retardant barriers for aircraft and other critical applications use very thin, thus light, sheet titanium. Heat exchangers, where the liquid or gas is aggressive, have titanium plates. Inert anodes in electroplating, and containers in the food and chemical industries also make some use of the metal. A small amount of titanium is used for jewellery and ornamental plaques. There is no doubt that there would be a dramatic increase in the use of titanium if the price could be reduced.

The high cost of obtaining titanium from its ore to a large extent inhibits the use of titanium and its alloys, which is compounded by the relatively low temperature range over which the materials are usable for long periods. It is therefore unlikely that use of titanium and its alloys will ever overtake the use of steel, but without doubt if the cost of producing titanium could be reduced significantly, then it would compete with many forms of stainless steels, in particular the austenitic type. If in addition a technically satisfactory, economical surface treatment were found, there could be a significant increase in its use.

At present there is no indication that the cost of producing titanium will be dramatically reduced, and while anodizing is possible and increases its corrosion resistance, it must be constantly monitored for imperfections or damage.

Titanium and its alloys are used in industries where there is a high level of technical competence. This, allied to the high cost of the metal, means it is seldom abused or wrongly specified.

There are instances where for safety reasons it is specified, but where a more economical metal would normally be used, such as in a sea water heat exchanger with titanium divider plates, where 316 austenitic stainless steel would be the normal material.

One failure investigated was a titanium-clad electric heating element used in electroplating. This accidentally became anodic in an aggressive hot solution, resulting in pin-hole corrosion.

Titanium on its own is difficult to weld because of its affinity for oxygen. Inert gas welding is seldom satisfactory for high quality welding, which must use a high vacuum.

F2 Titanium alloys – non-heat-treatable

There is a range of alloys based on titanium with aluminium, tin or vanadium. The alloys are strengthened but are not heat-treatable and there is some reduction in corrosion resistance. They can accept cold work to increase their tensile properties.

These alloys in general are used in the same areas as pure titanium where their higher strength – 700–800 N/mm^1 proof strength, against 300–400 N/mm^1 – is useful.

F3 Titanium alloys – heat-treatable

There is also a range of titanium alloys which can be heat-treated, whose mechanical strength is comparable to that of the lower range of alloy steels and which has excellent corrosion resistance. These alloys are used for compressor blades in gas turbines and for pump components for aggressive materials.

G COBALT AND ITS ALLOYS

This is a modern metal which is finding increasing use as an alloy element, many of its alloys being used for surface coating.

Cobalt (Co) has similar properties to nickel, is more expensive and is one of the three ferromagnetic metals, with iron and nickel.

G1 99% cobalt

Cobalt is now available as castings, strips, bars and sheets. It has few uses at present except as an electrodeposit for corrosion protection at high temperatures and as a light reflector under arduous atmospheric conditions. It retains its magnetic properties at high temperatures.

G2 Cobalt alloys

Cobalt alloy castings are difficult to machine, thus expensive. They do, however, find use in aggressive corrosive conditions, where high temperatures are involved. The trade names of Stellite and Haynes generally, but not always, indicate cobalt alloys. Cobalt alloys have specific properties for use in the nuclear industry.

The hot working range of cobalt alloys is very small, thus very few wrought products are produced. Many of the cobalt-based alloys are used for facing by metal deposit techniques. This allows the designer to use the mechanical properties and low cost of alloy steel, with cobalt alloys deposited on the critical areas to resist aggressive environments.

There are now sophisticated cobalt alloys which can be deposited by electroplating and are used in the aircraft engineering and nuclear industries. Some can be co-deposited with PTFE powder to produce low friction, corrosion resistant coatings.

The nuclear, aircraft and to a lesser extent petrochemical industries have their own specialist investigators and the failure investigator will seldom meet with cobalt alloy failures. With the increasing use of the material more experience will be required and obtained as failures occur.

The most common use for cobalt alloys is as metal deposits. There is a series of alloys involved containing elements such as chromium, nickel, molybdenum, tungsten, titanium and iron.

These alloys are deposited generally by the flame welding process, that is, standard welding, where the welder deposits the cobalt alloy on the substrate using as little fusion as possible consistent with adhesion. Alternatively the alloys are deposited in the form of a metal spray.

There are many forms of this and the majority of investigations of cobalt alloy failures are involved with either the standard of the deposit or the adhesion of the deposit. These will be found by visual examination, macro-examination or NDT. The investigator will use metallurgy to identify the state of the fusion, oxidation or porosity. If this is the reason for failure then it is unlikely that chemical analysis or mechanical testing would be required.

If failure was by corrosion it will probably be necessary to carry out analysis of the deposit once the metallurgy has shown that there is no excessive oxide, and the adhesion is satisfactory. This would require information firstly on the metal specification applied and then on the conditions under which the component was operating. The investigator must decide whether the material used was satisfactory for the conditions, then whether the conditions were altered or the wrong material applied. Only if the investigator is satisfied that the use conditions were either correct or unknown would chemical analysis be justified. This might require use of the scanning electron microscope (SEM) to identify surface or intergranular contamination.

Cobalt alloys are chosen when impact is involved with hot corrosive gases or liquids. A typical use for cobalt alloy deposits is on valve seats where the exhaust gases are hot and corrosive and there is impact each time the valve closes. Other areas would be where hot corrosive materials require to be transferred from one area to another, when cobalt alloy lined ducting could be used.

In addition to having good hot corrosion resistance, these alloys are tough, can withstand impact, and have more than reasonable hot tensile properties. They therefore find use in modern types of machinery where corrosive conditions exist along with high temperatures and very often impact conditions.

In some cases cobalt alloys are specified in lieu of standard stainless steels, where any failure would result in very high cost from shut-down or subsequent damage.

H MAGNESIUM AND ITS ALLOYS

Magnesium (Mg) is the lightest metal commonly used, having a specific gravity of 1.75 compared to steel (7.8), aluminium (2.5) and tungsten (19.3).

Magnesium is extracted from sea water by first evaporating the sea water to obtain the salt, and then using electrolysis in the same manner as for aluminium. The economics of evaporating sea water obviously demands plenty of sunshine, which eliminates most of northern Europe.

Magnesium has attributes which are useful in engineering. It is lighter than other metals commonly used, and its alloys have significant strength. It also has the ability to absorb vibration. This makes it a useful metal under circumstances where low vibration is required or for reducing noise levels.

Unfortunately, magnesium corrodes and is readily ignited, particularly in dust form; thus its use in the aircraft and similar industries has been severely limited. In cars and trucks, its use is confined to the unsprung components such as wheels on high performance models. It can present serious problems at machining, grinding and polishing, where its low ignition point results in a fire hazard.

Magnesium has a reputation of being readily corroded but there is no doubt that with the known procedures of chromating and painting correctly applied there should be no problem.

Because of certain drawbacks and its relatively high production cost, there is no increasing demand for magnesium or its alloys and thus the unit price remains high. It has been unfortunate in attracting dramatic headlines, such as where an aircraft ditched in the sea at take-off, and on recovery it was found that a large magnesium casting had 'disappeared' – by corrosion. Also, a racing car accident resulted in a magnesium casting igniting, with the 'fireball' killing a number of spectators.

Where weight is important such as in space exploration and where noise reduction is essential, magnesium still finds some uses. Pavement breakers employ magnesium 'skirts' to absorb some of the vibration, and thus reduce noise. It is no longer used in significant amounts in the aircraft industry, or in high performance motor cars with the exception of wheel castings.

H1 Magnesium alloys

There are a number of alloys available, with the 'rare earths', aluminium, zinc, manganese and zirconium being added to increase strength and castability.

The failure investigator would be faced with either corrosion or fracture problems. With corrosion it would be necessary to examine the standard of protection applied to identify any weakness, and to discuss the environment. Any fracture would be subjected to the same examination as for other metals.

Macro- and micro-examination could show casting defects, and it may be decided to carry out tensile testing. Chemical analysis for magnesium failures would be rare.

I TIN AND ITS ALLOYS

Tin is one of the 'ancient' metals, being the major constituent used after copper in bronze, and thus was known in the Bronze Age before the Iron Age. Tin is not difficult to win from its ore but is not as plentiful as lead. The Latin name for tin is 'stanium', hence the chemical symbol 'Sn'.

Tin is similar in many respects to lead, but invariably has better properties. Tin metal is non-toxic, easily worked, highly ductile and was thus used for drinking vessels, pipes, etc. in competition with lead. Many tin compounds are, however, highly toxic.

A point of interest is the fact that tin – much more valuable than lead – can be readily identified as it 'creaks' when bent. Thus anyone assessing water pipes in an old house does not require an analysis: lead pipes bend silently, tin pipes creak when bent!

A major use for tin was the production of tin plate for cans of all types. While some is still in use, tin plate has largely been replaced by aluminium and, to a lesser extent, plastic. Tin still finds a use in bronze bearings – this comes under the copper alloys.

I1

The only common use today for tin as a major constituent is in solder. Most common solders have approximately 60% tin, the remainder being lead with some minor elements. Where soldering problems are identified, analysis for tin and lead would probably be necessary.

J LEAD AND ITS ALLOYS

This is an 'ancient' metal known to Romans as 'plumbium', hence the chemical symbol 'Pb'. Lead could be easily obtained by smelting from rich ores. It has the considerable advantage that it can be joined at room temperature by forge welding and can be readily melted to produce intricate shaped castings.

From prehistoric times, lead utensils were used for eating and drinking, and lead culverts and pipes were used to convey water, and very often beer and wine. This continued from Roman times until the mid-20th century when it was found that lead in quite small amounts could cause problems to humans. It was identified that lead poisoning resulted in permanent mental retardation in children, and in adults a small amount in the bloodstream combined with alcohol caused aggression, and with larger amounts mental problems. There is evidence that the downfall of the Roman Empire was accelerated by the use of lead pewter drinking vessels!

In modern times it has been suggested that aggression in certain cities resulted from the combination of lead pipes and soft (low pH) water. Further investigation has shown that those involved in painting steel with lead-rich primers developed health problems. Serious problems were also identified when lead components or components painted with lead-based paints were welded or flame cut by workers who did not use positive breathing equipment. There is now evidence that the effects are exacerbated when those involved partake of alcohol. This situation contributed to the reduction in use of lead which is being replaced by copper and stainless steel, or plastic.

Use of lead in modern industry is almost entirely confined to the lead–acid battery. Modern batteries use a matrix which is lead plated. This reduces the weight and the amount of lead used.

Lead is used to a limited extent for lining vessels in the chemical industry as the metal has excellent corrosion resistance against the majority of chemicals involved in industry. Here it is normally alloyed with a small smount of antimony, which hardens the lead without seriously affecting the corrosion resistance. This use is being replaced mainly by plastics.

Lead sheeting which was commonly used on roofing as flashings is still used to a limited extent as it is easily worked and almost indestructible. This has been largely replaced either by zinc, which is more brittle and also has a limited life, or more commonly plastic, and zinc plated or galvanized mild steel.

J1 Lead alloys

Lead is still used alloyed with tin and certain other elements in solders. Tin–lead soldering is common in joining copper and copper alloy components in the electronics industry. Solder problems may require chemical analysis to verify composition.

Lead alloy bearings are still used to a limited extent but this is less common as the overall reduction in the use of lead together with Health and Safety measures means that the unit cost has increased.

Lead is almost invariably alloyed with tin for making light duty bearings which firstly are not capable of withstanding heavy duty loads, and secondly tend to wipe, that is to smear, at a lower load level than either tin or bronze bearings. The failure investigator will have identified the mechanism of failure, which could be expected to be creep, ductile tensile or brittle fracture. Very seldom will corrosion be involved.

Chemical analysis would be required only if full specification information was available for comparison. Micro-examination and simple mechanical tests would supply more information than analysis.

K TUNGSTEN AND ITS ALLOYS

The ore of tungsten is wolfram, and this is the origin of the symbol ‘W’. Tungsten is difficult to win from the ore which itself is quite rare and thus is expensive to process. The use of tungsten is largely limited to applications involving its high density of over 19, compared to steel (7.8), titanium (4.5), aluminium (2.7) and magnesium (1.7). Balancing weights are made from pure tungsten.

Tungsten metal is still also used as the filament for electric lamps. The wire is commonly coated with thorium oxide (thoria) to prevent sagging.

K1 Tungsten alloys

Tungsten carbides are the most common alloy of the metal. These have very high abrasion resistance, that is, they are extremely hard, are capable of resisting high temperature and are used for cutting tools. On the whole they do not have the ability to withstand impact and thus interrupted cuts should not use tungsten carbide tooling.

The tungsten carbides are invariably used in a matrix and the variation in the properties of the matrix material, that is porosity, analysis of material, etc., are of equal or greater significance than the tungsten carbides.

Where investigation is required micro-examination would identify the particle size, distribution, presence of porosity, etc., for comparison with any specification. If failure could be seen to be micro-cracking of the matrix material, chemical analysis might be required. Experience shows, however, that these materials are seldom submitted for investigation.

NON-METALLIC MATERIALS

The non-metallic materials used by industry can be given a variety of classifications and it is probably impossible to make a list which would be generally acceptable. The system used here is on the same basis as that for the metals, that is, to take an element, and then the compounds formed with that element.

There are only two elements of any significance with non-metallics: carbon and silicon. These are involved with natural materials where they are used with little alteration, and also with the same natural materials but with significant man-made changes. To complicate matters there is also a wide range of wholly man-made materials used in modern industry and particularly with the 'composites'. Without doubt there are now more man-made organic compounds than the total of metal alloys.

It will be appreciated that any comments made on non-metals in this book must of necessity be of a shallow nature compared with the information on metals, but could nevertheless be useful for comparative purposes.

There are five groups of non-metallic materials defined here as used in industry.

The metallurgist may be involved in problems at the initial stage but it would be more likely that a major failure would be looked at by the non-metallurgical specialist. The investigator must firstly have the same ability to see the mechanism of failure, and then the logic and technical capability to either carry out the technical investigation or to identify the correct technical source.

The groups listed and identified in approximate descending order of use by weight/volume are:

- L • Silicates – silicon based
- L1 • Stone and aggregates
- L2 • Concrete, cement, mortar, bricks
- M • Silicates – silicon-based man-made materials
- M1 • Glass
- M2 • Ceramics
- M3 • Refractories
- N • Carbon products
- N1 • Wood
- N2 • Wood products
- O • Plastics/polymers
- O1 • Thermoplastic materials
- O2 • Thermosetting materials
- P • Natural materials: wool, leather, etc.

L SILICATES

L1 Stone and aggregates

This is the natural stone material quarried, cut, shaped, crushed, etc. Stone was formed firstly as the Earth cooled, then by the abrasive action of rivers, glaciers, and compaction by pressure.

There are numerous types ranging from hard granite to sandstone. The majority are silica-based, but marbles are carbonates. They are largely used in compression mode, having relatively low tensile strength through faults in their structure.

Using the arch principal it has been possible, with considerable skill, labour and expense, to build impressive structures with limited height from stone. Modern buildings and structures use steel for the skeleton, clad with all forms of metallic and non-metallic materials.

Failure of stone structures is quite rare apart from earthquakes and man-made damage. Corrosion of certain stones, in particular sandstone, is common. The high sulphur/carbon dioxide, low pH acid rain of the Industrial Revolution era has caused considerable damage, in particular to industrial areas.

There has been much research and development work on cures for the problem. At the time of writing, there is considerable information on the cause but limited agreement on the long term cure. The environment in traditional industrial towns is now acceptable as the acid rain problem appears to be transferring itself from the doorstep of the producer to more distant neighbours. There no longer exists the 'smog' of recent years which held the contaminants at chimney-height; now instead the combustion products enter the upper atmosphere.

Aggregates can be produced from quarried rock/stone which is then crushed, or from river beds or material subjected to natural abrasion by earthquakes, glaciers or the effects of temperature change. Aggregates used for foundations and concrete mixes require specific control of size, shape and type of material, and is a specialist area where competent advice is essential. The geologist would replace the metallurgist for any failure investigation of stonework and compositions made with stones.

L2 Concrete, cement, mortar, bricks

These are man-made materials using silicates, sand, quartz, etc.

L2a Cement/concrete

Cement is a complex hydrated silicate which, when mixed with water, forms a semi-solid slurry which sets hard in 10–24 hours. Concrete always has sand and aggregate added, depending on the required application. The resultant material, correctly mixed and 'cured', has high compressive but limited tensile strength. This is a true composite.

Concrete/mortar is the most common medium used in conjunction with stone, bricks, etc. to fix them in position. The argument remains as to whether this is to keep them separate, or glue them together! Concrete is also used as pre-cast slabs, beams, etc., which may or may not be reinforced. These are then positioned in the same manner as stone and bricks. Pre-cast bricks or blocks without reinforcement are more commonly used for paving or partitions rather than for structural strength.

There are well-documented problems with concrete regarding the ratio of cement, sand, aggregate and water, efficiency of mechanical mixing, contamination and frost damage. Even so, it can be difficult to identify the reason for failure. Improper mixing, frost damage and contamination require specialist equipment and experience to pinpoint the cause. The ratios of sand, cement and aggregate can be identified by macro-/micro-examination, which requires considerable patience, skill and experience and careful sampling.

Chemical analysis, except for specific substances, contributes little to problem solving. High alumina cement can be identified by analysis. This was an additive to prevent frost damage, which was found to cause shrinkage of the finished product.

Concrete is a composite in its own right. Reinforced concrete is a classic composite. This uses the tensile strength of steel with the compressive strength of concrete to give a valuable building material, allowing the construction of high-rise buildings. The steel bars are first fitted, and where necessary moulds constructed, then the concrete poured and very often vibrated to obviate porosity. Pre-cast beams can be mass produced with the reinforcing bar protruding where necessary for joining.

A development now in common use is to pre-stress the steel reinforcing bars. With this the steel bars, wires or strands have a tensile load applied to approximately the yield strength. With the load applied the concrete mix is poured, and the tensile load retained until the mix has set. This is almost always a factory-made product. Site pre-stressed concrete products can, however, be achieved by having the beam cast with tubular inserts. When the concrete has set, the cables are passed through the tubes and the tensile stress applied. With the load on, concrete is forced into the tubes from each end and when this sets, the tensile force is removed, thus putting the tubes into compression.

Failure of these pre-stressed units should involve a metallurgical investigation as there is always the possibility that fracture of one or more of the steel reinforcement bars would remove or reduce the imposed compressive stress.

Concrete failure can be a result of fatigue (tensile stress), compression, corrosion (contamination) or abrasion. It is unlikely to be heat-induced. The mechanism can usually be seen by the investigator, but it is more difficult to identify the 'Why' with these materials than with metals.

A common problem is corrosion of the reinforcing bar, and this can be very serious with pre-stressed concrete. Concrete, however, is alkaline and thus tends to inhibit steel corrosion. Corrosion/contamination of concrete can result in various problems. Almost all acids attack cement. The aggregate used for high quality concrete must be free of contaminants. Soluble sulphate is normally unacceptable: significant amounts of sulphate can seriously affect the properties of concrete. Aggregates used must therefore be controlled for sulphate content. Sulphate can be water soluble when it can affect the properties of the concrete in the short term, acid soluble when the effect will not be found for some time, or insoluble in which case it does not affect the concrete.

The location of the steel bar with respect to the surface can be critical. Too thin a surface layer can cause exfoliation, while excessive thickness between the outer steel bar and the surface can result in brittle failure of the surface concrete.

L2b Bricks, tiles

These are composite materials traditionally based on clays.

Clays have a particle size of less than 0.005 mm (0.5 microns), are called colloids, and show considerable variation in their composition. Some contain organic matter, including coal dust. These could be used for self-firing bricks if they also contained fusible material such as silicates.

Traditionally bricks were pre-formed and then fired to fuse the lowest melting point constituent(s). This resulted in a variable appearance and also technical problems. Such bricks are fired in kilns, and self-firing bricks are more economical than furnace-fired bricks.

Most modern bricks are manufactured to stringent quality standards. This means that when problems are identified, the product being investigated can be compared with and tested against a specification. Bricks are now mass-produced from silicate clays with additives to give strict control of colour and size. Modern bricks have holes to ensure heat penetration and to give homogeneous fusion, and thus stability. They are fired in temperature-controlled furnaces or kilns.

Most of the above comments also apply to components such as glazed tiles for kitchens, floors and bathrooms, and unglazed roofing tiles. The glazing is from fused silica, applied as a paste prior to firing. There can thus be two stages of fusion, for example certain bricks could be made as described above, then after cooling have the glaze applied as a wash which is dried and then fused by heating.

M SILICATES - SILICON-BASED MAN-MADE MATERIALS

M1 Glass

Glass is still used in greater quantities by weight than plastics/polymers.

Glass basically is fused silica with additives for various reasons of application. Good quality glass has high tensile strength equal to, or better than, the strongest metal, but very poor impact strength and ductility for any except the most specialist of components. Glass, therefore, is expected to break when subjected to impact or bending. It would therefore be most unusual for any normal glass product ever to be submitted for chemical analysis or mechanical tests because of mechanical failure. Where, however, glass is used for other purposes, for example refraction, reflection or transmission of light, then if these parameters did not come up to the standard required, chemical analysis would be necessary to identify the impurity causing the problem.

For example, a very few parts per million of iron or copper will colour glass quite dramatically. These and other additives can be used for decorative purposes, for example to produce stained glass or imitation jewellery, but also can act as unwanted impurities where optical or other properties are required.

Glass is now used in fibre form as an insulating material, or as part of a plastic composite. When composite failures are being examined then the quantity of glass relative to the plastic/polymer would be of significance, as would the fibre size, the adhesive and the method of lay-up. It would be unusual that the quality of the glass itself would be suspect.

Macro- and micro-examinations would supply significant information which might indicate that specialist help would be required if this was economically justified.

Where glass fibre is used for insulation, few problems should be found, apart from moisture or other obvious contamination. The glass fibres, apart from quantity, size and distribution, would seldom be at fault. Contamination of the fibres can result in a dramatic reduction of the insulation property. When the contaminated material is moist, fungal growth can occur which then increases the problem.

Fibre optics are being developed at a rapid pace for a number of applications, none of which can be expected to involve the general failure investigator except in exceptional circumstances. These are high technology products installed and applied under controlled conditions where material problems would be unusual, and when present would be investigated by in-house specialists.

M2 Ceramics

This covers a wide range of materials, generally clay-like (that is colloidal) in the natural state, which have been fired at high temperatures.

In the general sense this covers materials such as glass, porcelain, cement and abrasives (silica, but also alumina and carbides, etc.), but it usually refers to certain bricks and tiles. Cement, bricks and glass are covered in sections L2 and M1, the refractory materials in M3.

The properties often expected from ceramics would be high electrical resistance and high tensile/compressive strength. Ceramics suffer from very low impact and ductility, low heat conductivity, and low expansion.

Ceramics are used extensively in the electronics and electrical fields, in addition to decorative tiling and ceramic ware.

It would be very seldom that the metallurgical investigator would be involved in examining failure of pottery, porcelain vases or dinner sets. These can be expected to have cracks where the fused silica glaze (porcelain) has a different expansion to the underlying pottery (fused clay). Fracture by impact from dropping or damage through careless use of the dishwasher are accepted as routine and would not be investigated.

There could be occasions when failure of a ceramic assembly was suspect and investigation required. This would involve examination of the joint to identify if failure was from the adhesive. This would be visual, and perhaps involve micro-examination which would require destruction.

Ceramics used in industry are not normally subjected to mechanical stresses. These are highly specialist components where failure for electrical or electronic reasons would be investigated by the specialist. Where mechanical failure, such as the bond of a metallic component to a ceramic, is involved, visual, macro-, and where necessary micro-examination might be required.

Ceramics will almost always have lower expansion characteristics than a metal, thus heating of the metal could cause problems. This might be identified by micro-examination showing grain growth in the metal, possibly with other evidence such as oxidation. The metallurgist would require to co-operate with the other specialists to obtain an answer. This would seldom require chemical analysis or mechanical testing.

M3 Refractories

These are materials not deformed or damaged by high temperatures. It is generally expected that the temperatures will be in excess of 1000–1500°C. The term refractory refers to this property, not to any specific material.

Many metallic compounds such as alumina (Al_2O_3), magnesite ($MgCO_3$), dolomite, thoria and zirconia are used in addition to silica (SiO_2) and silicon carbides (SiC).

Refractory materials generally have low tensile strength, little or no ductility or impact resistance, but good compressive strength, and are not oxidizable.

Failure in most instances can be expected to be by contamination or excessive heat. Some processes which use refractories are more efficient at high temperatures, thus there is a tendency to operate as close to the maximum as possible, and any error can result in failure.

Contamination of a refractory material will result in a lowering of the fusion temperature at the contaminated surface with subsequent failure. The investigator will thus look for contamination and may require chemical analysis. He or she should be aware of the possible contaminants, and if these are absent could assume excessive temperature to be the cause.

Spalling – that is, exfoliation or flaking of the surface material – can be caused by a rapid increase in temperature. Refractories will always have low thermal conductivity. Thus the heat will not be conducted quickly away from the surface, and although refractories almost always have low expansion characteristics, there will be some expansion. There will thus be shear stress between the heated surface and underlying material. Spalling is often the result of cyclic heating. Some information on the degree of heating can be obtained from macro-/micro-examination, which will show if any fusion has occurred and may indicate cyclic heating.

Personnel experienced with refractory materials can identify from visual examination alone considerable information, before requiring chemical analysis, or other physical checking.

N CARBON PRODUCTS

N1 Wood

Wood is a natural product and has specific but variable properties depending on the type. One which is common to all wood is that its tensile properties are related to its length and not cross-sectional area. In other words, this can be likened to a book which when pulled along its length will have reasonable strength, but when the pages are opened will have poor strength. Different types of trees can have better through thickness strength but there is no wood where the transverse strength equals the longitudinal strength.

Wood is used in a variety of products, the most obvious being in the construction industry, where rafters, windows, doors and flooring are made of wood or wood products. Furniture and cabinets for all manner of equipment were at one time produced from wood. Many of these are now made either in plastics/polymers or from fibreboard or composites. In many cases it can be difficult to tell the difference from real wood.

Wood failures are normally from rot or tension, generally bending, rot being grouped under corrosion/contamination/degradation in this book.

Where rotting has occurred and the life of the product has been short, then there is an argument for carrying out an analysis to find the contamination. If, however, it is found that the wood product has been kept wet or under very humid conditions for long periods, then this would be the cause of the problem, although there are some woods, particularly hard woods, which should not rot even under these conditions.

Most woods, particularly soft woods, require the surface to be sealed by painting or some other treatment. Where this fails and rot sets in, re-painting may seal the surface and stop the rot, but commonly it will continue, causing paint failure and continuing problems unless the dampness is eliminated and all fungi and rotten wood removed.

There are treatments, which often require specialist advice, to kill the fungi and bacteria in rotting wood. These may require specific coatings which are compatible with the original problem and the treatment chemicals to be successful in the long term.

It is not often appreciated that there are many buildings hundreds of years old either of wooden construction or having major wooden construction elements such as beams. Very often these have had little or no maintenance and present few problems.

While fire is a serious hazard with most wood, there are now fire-resistant treatments and it is in fact quite difficult to ignite most mature woods of large cross-section.

N2 Wood products

These include plywood, fibreboard, paper and other cellulose products.

Plywood

This is a composite of thin layers of wood with the layers glued together at right angles to each other, thus eliminating any directional problem associated with timber grain.

Failures are generally related to the adhesive, excessive heat or moisture. Delamination is the most common problem. As many plywoods are low cost items, investigation is not usual. The most common reason for failure is that cut edges are not protected from moisture.

Fibreboard is made from chippings, cuttings or fibres from wood, which can be of high quality timber, scrap, waste or trimmings. The raw material can be of random size or highly controlled tiny fibres. The type of timber used, moisture content, etc., may or may not be controlled. The type of glue, control of mixing, temperature and pressure will all have a significant effect on the finished product.

It is now possible to produce board products from wood 'wool' which have high strength in all directions, and good appearance. These use specific types of timber, with fibres produced under highly controlled conditions regarding diameter, length, and moisture content. They are then mixed with thermosetting glue and subjected to heat and pressure under controlled conditions. All these aspects have to be taken into consideration by anyone investigating fibreboard problems.

The investigator would use visual examination to identify if further investigation would be economically justified. This would take into consideration whether or not there was a specification available, and the quality standard involved. Only if the cost of further investigation such as macro-examination, tensile testing and chemical analysis appear justified should the investigation proceed.

Paper

This is a major product but will seldom be involved with an investigation outside the production companies.

Composites

With composites, if the properties are suspect then there are specialist techniques for examining these firstly mechanically and visually, including the particle size and the type of glue used. This would come under micro- or macro-examination.

Where no obvious reason for the failure is identified, chemical analysis might be necessary. This would require firstly information from the specialist who produced the product regarding the parameters to be analysed, and then involve an independent analyst with the necessary equipment and experience to carry out an analysis and translate the results.

Wood and its products, however, are seldom submitted for chemical analysis at failure investigation.

O PLASTICS/POLYMERS

Polymers are compounds consisting of large molecules built up by chemical combination of monomers (small molecules) into long chains. In the main the 'back bone' of the chain is based on the carbon atom, with many other elements involved in the 'ribs, arms and legs'. There are now a series of components where the back bone of the polymer has the element silicon (Si). These almost invariably have at least one of the ribs, arms or legs with carbon. These materials are known as 'silicones' and are included in this group of plastics/polymers.

Plastics/polymers to some extent are the up-and-coming materials in engineering, but they will rarely compete with metals where strength, impact resistance, ductility and temperature are involved.

Note that the word 'plastic' to the metallurgist means 'readily deformed': hot metals are subject to 'plastic deformation'. To others the word 'plastic' is a man-made material which can sometimes be brittle, and difficult to deform. The industry in general defines plastic as a material produced by the polymerization of monomers. There are a variety of other definitions or understandings of plastic. This book always uses the term 'plastics/polymers', and includes in this materials such as rubbers, natural and synthetic, and silicones.

Plastics/polymers are therefore mainly carbon-based but include silicones. The carbon-based materials have a temperature limitation of around 200°C, which restricts their usefulness. Specialist plastics are usually expensive, but soften or decompose at a higher temperature. However, no plastic/polymer will normally be found in service above 300°C except for very special applications. Some silicones are stable to slightly higher temperatures, but these are generally in liquid form.

A feature with plastics/polymers is that they are produced as relatively pure chemical compounds. This can be likened to the production of aluminium or aluminium alloys, steel, etc., which are then submitted to the engineer for development purposes. The plastic/polymer can have its properties improved (or worsened) by additives. These additives are sometimes made during the forming operation of moulding, etc., and can alter a product's properties to varying degrees.

The mechanisms of mechanical failure with plastics/polymers will in most cases be not unlike those for metals, that is, tensile, brittle, ductile, fatigue, and is more common than is often acknowledged. Torque/shear and compression failures are not uncommon.

Many plastics/polymers are truly plastic, some being liquid or semi-liquid; thus 'fracture' takes place under zero or very low tensile stress.

Chemical analysis of a specialist nature can be carried out on these materials but it must be emphasized that the results will require to be interpreted by someone with expertise and supplied with the information on the problem and the intended specification. The specialist must be capable of interpreting the highly complex information obtained from the various forms of analysis. These will include infra-red, where a 'fingerprint' is obtained of the composition of the material. This can be compared with the 'fingerprint' of the original material. This can inform the expert investigator whether or not the material has altered during its manufacturing or service life. It requires special equipment, considerable experience, and can be confused by the 'extenders' added during manufacture.

The polymers/plastics are one of the major components in the up-and-coming composite materials. One of the problems which has been shown to exist is that the other component in the composite which might be a metal, carbon or glass fibre will have a fixed composition which will not alter its properties, whereas the plastic/polymer can have its properties altered, firstly by the mixing and secondly by variations in the temperature and pressure used during forming operations.

Finally, there is the problem that plastics/polymers unlike metals have a tendency to become older, like human beings. Metals do not age except in very specific and well identified circumstances. They might corrode, but this is a visual and readily identifiable surface effect. Metals do not normally change their properties with time at ambient temperature, as do plastics. Plastics/polymers are susceptible to both temperature and time, and also to certain types of light which will accelerate the ageing process, almost invariably resulting in the plastic/polymer becoming brittle. This applies more to the organic (carbon-based) polymers than the silicones.

The ageing effect can be identified by specific forms of specialist chemical analysis which will show that the molecular bonding has altered.

Plastic coatings

Included in this group are corrosion protection coatings. These come under the generic title of 'paint'. In some cases these are straightforward plastics/polymers applied in liquid form, sometimes with a solvent. When heated, polymerization takes place and the product changes to become a semi-solid plastic coating.

Most paints have three components:

- the vehicle
- pigment
- thinners

The traditional paint was linseed oil (the vehicle) with a solid pigment normally metallic oxide, and almost always thinned with turpentine. This was applied as a liquid, with the linseed oil oxidizing to a solid to give the dried film and the turpentine thinner evaporating. This is an early example of a true plastic/polymer.

Modern paints are complex chemicals of various types, with the same three components, a vehicle, pigment and solvent.

The vehicle can be one of several organic chemicals which converts chemically from a liquid to a solid, sometimes requiring heat (stoving). The pigment is commonly a stable non-toxic organic solid, with the thinner commonly being MEK (methylethylketone). Metallic particles and oxides tend only to be used for specialist paints such as for active (galvanic) corrosion protection.

There is now overwhelming evidence that paint failures are the result of faulty preparation (75%) or faulty application (20%), and very seldom the paint itself. Modern quality control and the development of specialist paints have dramatically improved the situation. There are plastic/polymer coatings, some quite complex, which can accept considerable abuse, but the application of quality control to the preparation techniques and better control at application have achieved more than the development of improved plastic/polymer composition.

Some specialist primer paints have active ingredients which can convert rust – iron oxide – and are specified for corroded components. There is a history of problems with these paints. The investigator has the added difficulty of identifying if the active ingredient did not work, or the application was wrong, or there was some other, unknown, reason for failure. It would appear to be sensible advice that these active paints should be treated with caution, and that the conventional recommendation of removing all contamination before applying the primer should be followed.

The investigator using visual examination and micro-examination alone will identify in most instances the cause of failure without chemical analysis. Some corrosion protection failures are the result of contamination, chemical attack or atmospheric attack on the plastic/polymer coating, and investigation into paint or coating failure might require chemical analysis of the contamination, but seldom of the coating itself.

Macro- and micro-information will, however, supply the details regarding whether failure is at the surface/coating interface, or at the coating surface working towards the interface. Only if the coating surface was attacked would the coating itself be suspect, either from contamination or from being of poor quality.

In some cases the paints are two-pack which are mixed by the painter. Unless this is properly carried out the paint will not harden, or the colour will not be stable. This problem can be identified by visual examination but may require chemical analysis with sophisticated equipment, and specialist interpretation of the result.

O1 Thermoplastic materials

These materials more nearly conform to the metallurgical understanding of the word 'plastic' than those in the thermosetting group. Many, but not all, of the materials in the thermoplastic group have high ductility and softness. The unique characteristic, however, is that the materials, by heating, pressure or solution, can be remoulded, whereas the thermosetting group cannot.

It is not possible in most instances to identify by visual or tactile means whether a plastic/polymer is thermoplastic or thermosetting.

Products made from thermoplastic sheet and fibres include packaging materials, electrical insulation, clothing, corrosion protection coatings and artificial leathers. Mouldings include light fittings, car fittings, piano keys and some panelling.

The materials in this group include nylon, cellulose derivatives, polyethylene, polypropylene, vinyl acetates, acrylics, polytetrafluoroethylene (PTFE) and styrene.

O2 Thermosetting materials

These are generally produced in powder form which, when heated and injection moulded, sets hard and would be destroyed if reheated.

P NATURAL MATERIALS: WOOL, LEATHER, ETC.

Wool, cotton and leather are still used in large quantities in clothing, carpets, furnishings, and other forms of textiles. They are under pressure from plastics/polymers such as nylon and polypropylene, for which the abrasion resistance has been shown to be equal and the tensile strength can be better than for wool, cotton and leather.

There is now little or no demand for natural fibres such as silks.

With wool there are considerable differences in characteristics such as fibre diameter, length and tensile strength. There can also be differences depending on the weather and health of the sheep involved, when the fleece was shorn, and how it was processed.

Wool, cotton and to a lesser extent leather have a considerable advantage over most plastics in that they can be dyed, that is, the colour is absorbed into the fibre and not just the surface.

It is often difficult to identify the difference between the natural and the man-made fibre or sheet materials by visual examination.

At present there are few instances where failure of fibres or sheet material is subjected to investigation by investigators outside the industry or company producing the product.

There are considerable skills, many of a traditional nature, involved in the initial production, handling, treating, then pressing of these materials, and only in unusual circumstances would an independent investigator be involved in failure analysis.

Heat treatment of metals | 14

The term 'heat treatment' can be used to cover a multitude of operations, all involving heating metal components in a furnace, by induction, radiation, infra-red or flame. There are also various ancillary operations involved where metals are incidentally 'heat treated' during welding or through friction, etc.

While, without doubt, the term can and is used for non-metals, there does not appear to be any simple relationship between the names used and the processes involved, and thus no attempt has been made to include non-metal processes in this book.

INTRODUCTION

There are two important reasons why heat treatment of metals should be given a chapter in this book. Firstly, heat treatment of metals allows modern industry to operate in a much more efficient manner than it otherwise could. It is not possible to conceive drilling for oil where drill pipe over a mile long could be involved with high torque, high tensile loads. This would not be possible using wood, wrought iron, cast iron, aluminium, mild steel or any other material.

Aircraft could not fly efficiently without heat-treated aluminium alloys, and heat-treated copper–beryllium tools are essential in flammable or explosive atmospheres.

Secondly, heat treatment broadens the possible range of characteristics, but faulty heat treatment is a common reason for failure from lack of quality control or more usually accidental 'heat treatment' through over-heating.

It is the experience of the author that the heat treatment of metals is poorly understood, even by some metallurgists. The control parameters at welding are good examples of this lack of understanding. In order to control heat treatment at welding the two procedures used are firstly to pre-heat, and secondly to control the interpass temperature. Discussing the reasons for these with welding engineers, and in many cases metallurgists, it is clear that there is a lack of understanding of the interpass temperature. The reason for preheating is often not completely understood.

There are two main processes in use. In order to harden steel it is necessary to take it above a certain critical temperature and then to cool it rapidly. This results in the steel being hard and brittle. The steel is then tempered or 'drawn back' to reduce the tensile strength and brittleness and increase the ductility and toughness. It is the art or science of good heat treatment in these areas which allows steel to have high tensile strength with toughness.

With non-ferrous metals the heat treatment, including that for duplex stainless steels, is reversed. The initial treatment known as 'solution treatment', is to heat the metal above a certain critical temperature and then to quench it. The more rapid the cooling the more efficient the solution treatment will be, resulting in the material being in its softest, most ductile condition. This is followed by what was initially known as an 'ageing operation' – now termed 'precipitation treatment', the terms being synonomous. This results in hardening of the material by a complicated metallurgical process which reduces the ductility and increases the tensile strength.

There are other differences in the two treatments – for example, once steel is correctly tempered it can be taken to within 1 °C of the critical temperature for infinity and it will not alter its hardness.

Ageing/precipitation treatment, on the other hand, is a time/temperature procedure. That is, when a component is held at the same temperature for a longer time it will initially increase in tensile strength and once it reaches a peak at a certain time/temperature combination, the tensile properties will be reduced. Depending on the temperature and alloy, the component can become quite soft if held at the temperature for a long time.

The two procedures, that is, hardening with tempering and solution treating with precipitation treatment (ageing) the most important when mechanical properties require to be improved.

There are several case histories attributing poor heat treatment as the cause of failure.

Additionally it is possible that there are considerable sums of money wasted in the controls applied to post-weld heat treatment (PWHT). Very often the procedure is unnecessary and even when necessary, a sophisticated procedure is chosen which would not be required except under very specific conditions.

This chapter will attempt to describe with the minimum use of technical terms, the heat treatment firstly of steel, and secondly of non-ferrous metals which include the duplex stainless steels.

Steel heat treatment is discussed first, with all common terms defined, described and explained. With the non-ferrous alloys, including duplex stainless steels, then only solution treatment, precipitation treatment (ageing), and annealing will be described as these are the only processes of any relevance.

There is also information on the problems which can arise either with faulty control at heat treatment, or what is more commonly found at failure investigation, that some incidental feature has resulted in uncontrolled 'heat treatment'.

STEEL HEAT TREATMENT

Steel is an alloy of iron and iron carbide. It is commonly stated that steel is an alloy of iron and carbon but this is not true as unless the carbon is present as iron carbide, the necessary reactions will not take place. (Iron and carbon alloys are cast irons.) It is, however, accepted that the convention is to quote the carbon content as a percentage of the total steel content, knowing that the carbon will be combined with iron, as iron carbide or cementite.

Iron has two critical phases – ferrite or alpha iron, and austenite or gamma iron. Metallurgists love to make use of Greek and Latin prefixes but for the remainder of this section the Greek will be left out and reference made only to 'ferrite', 'austenite', 'iron carbide' and 'pearlite'.

To understand steel heat treatment it is essential to refer to the 'iron carbon diagram' (Figure 14.1).

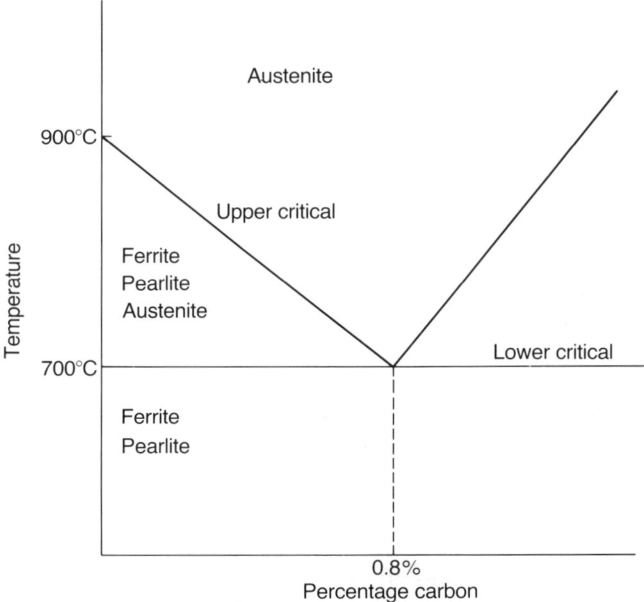

Figure 14.1 The iron/carbon phase diagram (simplified).

Figure 14.1 shows only a tiny portion of the complex phase system that exists with iron and carbon, but this portion relates to the vast majority of heat-treatable steels. (The diagram given is in a simplistic form, which may annoy the purist metallurgist, but it is believed that if the reader can appreciate that the important temperatures for steel heat treatment are approximately 700°C and 900°C, and that the significant carbon content varies from 0.1 to 0.8% carbon, this is better than being confused with accurate temperatures, which vary in any case depending on the steel composition.)

Like Topsy, metallurgy of steel 'just grewed' and in the process proliferated technical terms or generic names, many of which are duplicates or even quadruplicates.

This section will use only the terms:

- Ferrite
- Austenite
- Iron carbide
- Pearlite
- Martensite

Alternative names for these are:

Ferrite – alpha iron, body-centred cubic iron, bcc;
Austenite – gamma iron, face-centred cubic iron, fcc;
Iron carbide – cementite, Fe_3C;
Pearlite – a metallurgical structure made up of ferrite and iron carbide in layers;
Martensite – an unstable form of austenite plus iron carbide produced by quenching, which when heated transforms into 'tempered martensite' which has various forms known as 'troostite', 'sorbite' and 'bainite'.

The important facts to be derived from the phase diagram are:

- Ferrite cannot dissolve iron carbide;
- Austenite can dissolve iron carbide;
- Ferrite transforms into austenite, starting at a temperature of 700°C;
- Pure iron – ferrite transforms over the temperature range of 700 to 900°C into austenite;
- When pearlite is present along with ferrite, this will not affect the 700°C temperature at which the transformation of ferrite into austenite starts, but, as the phase diagram shows, the iron carbide, present in pearlite, reduces the upper temperature for the conversion.
- Within the confines of the upper and lower critical temperatures, and the 0.1% carbon to 0.8% carbon range, the ferrite is converted to austenite and the austenite dissolves the iron carbides;
- Above the upper critical temperature, all ferrite is converted to austenite and all the iron carbide is dissolved in the austenite.

These reactions all take place in the solid state. It is appreciated that it is difficult (even for metallurgists) to accept that solids can 'dissolve' solids. This in fact is no more complex a process than liquid tea/coffee dissolving solid sugar, or liquid whisky dissolving or being dissolved by water. The physical chemistry is similar whether solids, liquids or gases are involved, but the reactions between solids take longer than when liquids or gases alone are present.

The basic heat treatment process

It is the fact that iron has two phases, ferrite and austenite, one of which, austenite, can dissolve iron carbide, which makes steel unique in metallurgical terms. When a piece of 0.4% plain carbon steel is heated to above approximately 800°C, all the iron carbide (present as pearlite) will be dissolved in the austenite. If it is cooled slowly to room temperature it will revert to ferrite and pearlite, which can be seen in the micro-structure, which has approximately 50% pearlite and 50% ferrite grains. This is termed 'annealing', and will result in some grain growth.

If it is air cooled to allow homogeneous heat loss at a reasonable rate to 700°C – a process termed 'normalizing' – the grain size will be controlled and the structure will again be 50% ferrite and 50% pearlite, but with a smaller grain size.

If, however, the steel is cooled rapidly – that is quenched – from above 800°C, the iron carbide cannot revert to pearlite and is retained in solution in the austenite in an unstable state. This is termed 'hardening'. The micro-structure will show a homogeneous needle-like (acicular) structure, known to the metallurgist as martensite. This will be hard and brittle, with the hardness being related to the iron carbide (carbon) content. Various micrographs are shown in Figure 14.2 illustrating these points.

If the component is heated to 100–150°C there will be no visible change to this micro-structure, but it is known that alpha martensite is converted to beta martensite. This improves the ductility by a small extent without affecting the hardness.

If the steel is then heated above 200°C the needle-like hard structure will change, the needles becoming less sharp. Above approximately 400°C the needles will disappear to give an indeterminate structure – becoming like 'gingerbread' – and as the temperature increases, spheroids will appear, gradually changing to the granular structure of ferrite plus pearlite when held at 650–700°C for approximately two hours.

There is a drop in hardness (tensile strength), with an increase in ductility proportional to the increase in temperature. There will normally (but not always) also be an improvement in impact properties. This heating between 200°C and 650°C is 'tempering', with the fully tempered structure being not unlike that of the normalized or annealed condition.

The above is the basis of the 'heat treatment' of plain carbon steel.

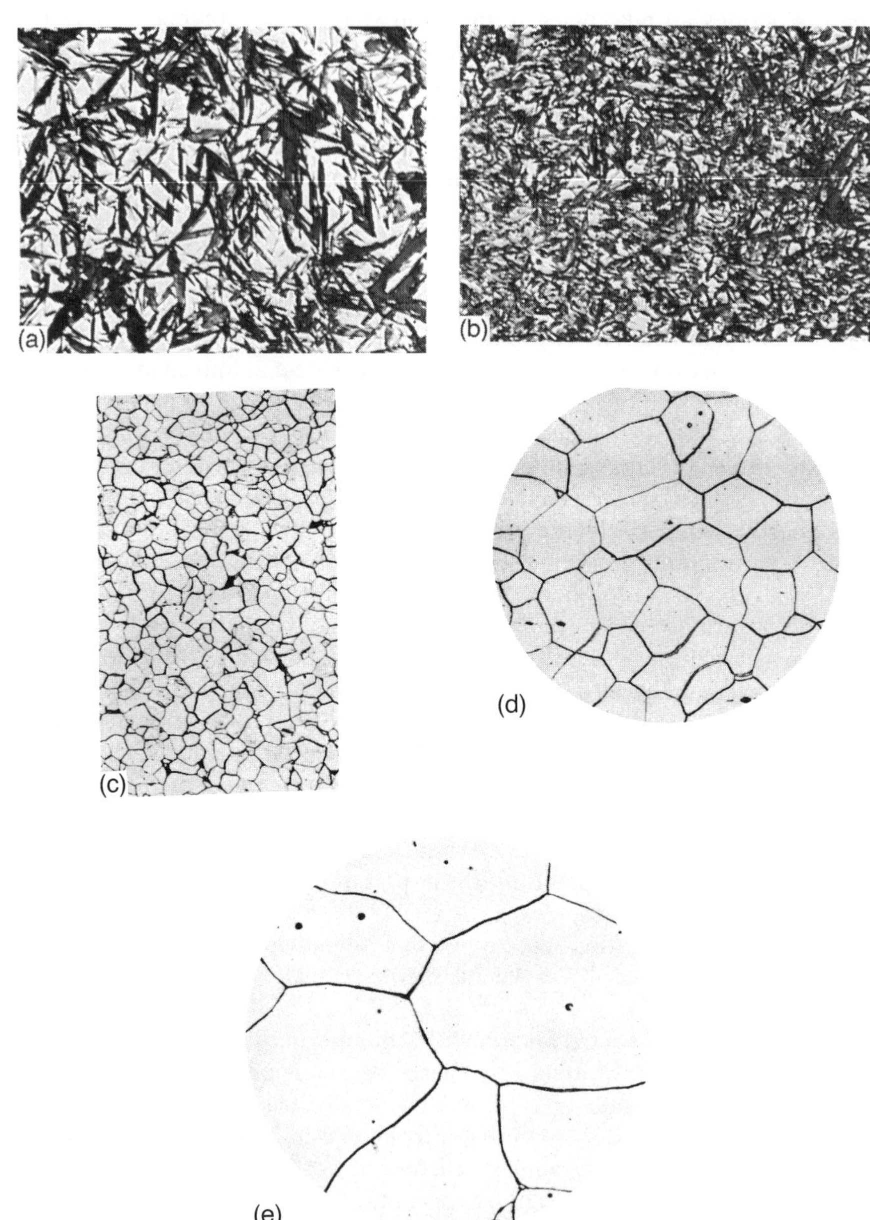
(a)
(b)
(c)
(d)
(e)

It must be appreciated, however, that there are other factors which the heat treater must understand and control.

Factors influencing heat treatment of steel

The first of these is that at approximately 700°C the phenomenon of recrystallization starts. This results in the existing grains being destroyed and infinitely small seed grains forming which immediately start to grow. This grain growth is time/temperature related. Thus the higher the temperature and longer the time at that temperature, the larger the grain.

The importance of grain size is discussed and illustrated in Chapter 8 on Micro-examination.

Basically the problem is that impurities agglomerate at grain boundaries. Thus a small grain will have a thinner layer of impurities than a large grain, if the total impurity count is the same for both structures. Thus large grained steel will almost always be more brittle than fine grained steel. This presents the heat treater with three possible problems:

1. Too low a temperature or too short a time at the correct temperature means that all the carbides will not be dissolved in the austenite, thus the maximum hardness will not be achieved, though the steel will be ductile.
2. Too high a temperature, or too long at the correct temperature, may give the correct hardness but will result in grain growth, and thus brittleness.
3. Too long at too low a temperature can also give a brittle structure and one which is low in tensile strength (hardness) because of grain growth.

The above has confined the discussion to plain carbon steels, which have low hardenability as the iron carbide which forms the pearlite with ferrite is a 'fit, healthy and active' constituent. This means that it is quite readily dissolved in austenite, but also that it escapes equally readily.

It should be appreciated that any object heated can absorb and lose heat only via the surface. Thus a rectangular component which is thin will heat and cool more rapidly than a thick component.

If the component is likened to a book, then the word in the centre of the middle page must reach the temperature for the carbide to be taken into solution

Figure 14.2 Grain structures of carbon steel after different heating and quenching treatments. All specimens etched in nitric acid/alcohol (nital). Magnification ×500. (a) 1% carbon steel quenched from too high a temperature, showing large needle-like structure which is very brittle; (b) 1% carbon steel correctly quenched, fine grained, not as brittle as (a) but just as hard; (c) 0.05% carbon steel in the correctly treated condition with fine grains;(d) 0.05% carbon steel cooled more slowly and held for a longer time or taken to too high a temperature; (e) 0.05% carbon steel which has been overheated, that is, held for too long a time at too high a temperature.

in the austenite. To be hardened the heat must be removed from that word fast enough to prevent the carbide escaping from solution. It will thus be seen that thick sections are more difficult to harden than thin sections.

The Victorians and Edwardians were confined to plain carbon steels and developed quenching techniques to remove the heat. These included agitation and using brine instead of water, and tradition has it that the urine of red-haired youths gave the best results! For all this, sections greater than 15–20 mm could not be fully hardened until alloy steel was discovered.

With quite small amounts (0.2–2%) of elements such as chromium, molybdenum and vanadium, all of which form carbides preferentially to iron, it was found that the reactions were slowed. It takes longer for these carbides to be dissolved in the austenite, as these carbides are not as young, healthy or active as the iron carbides, and cannot so readily escape from solution in the austenite at cooling; thus larger sections can be hardened. Other complications then arise for the heat treater.

Some carbides require a longer time at temperature to be dissolved, others such as chromium raise the upper critical temperature, so thus require a higher temperature. This increases the possibility of grain growth.

The increased hardenability by itself can cause problems. When austenite converts to martensite there is an increase in volume of approximately 4%. This increase in size is rapid, while the normal change in volume because of the coefficient of expansion takes place over a range of temperature. Considerable stress can be involved at this point of change, and it is important to avoid stress raisers such as lack of radii at sudden changes of section.

The effect of the problem of 'tramp' elements on hardenability, and the importance of the carbon equivalent must also be appreciated. The carbon equivalent (CE) has been discussed elsewhere, being a comparatively modern idea introduced for control of welding, to estimate the degree of hardenability using a formula:

$$\text{CE\%} = \text{C\%} + \frac{\text{Mn\%}}{6} + \frac{\text{Cr} + \text{Mo} + \text{V\%}}{5} + \frac{\text{Ni} + \text{Cu\%}}{15}$$

It is accepted that a steel with a CE of less than 0.4–0.45% will have limited hardenability, requiring water quenching to produce martensite. Thus steel with a CE of less than 0.4% should not cause serious problems at welding or through accidental heating.

With a CE above 0.45% and below 0.7%, martensite can be produced with less rapid cooling, thus oil quenching would be used for hardening of small section components. While water quenching would be necessary for larger sections, components must not have any rapid change in section or sharp radii. Any welding would require pre-heating, and control of heat input or interpass temperature control.

With a CE of 0.6 to 0.8% there would be excellent hardenability, oil quenching achieving a martensitic structure in all reasonable sized components. Care would be required to eliminate all stress raisers. Welding is possible, but careful pre-heating and control of heat input is essential, with post-weld heat treatment advisable on high integrity components to remove any local excess hardness.

With CEs of 0.7–1.0%, there would be high hardenability, with sections under 8–10 mm hardening in air. Welding would require great care, careful pre-heating, low heat input, post-weld heat treatment and a crack test.

With CEs above 1.2%, hardenability is excellent but with the possible problem of cracking at hardening. Welding would not be recommended except under very carefully controlled conditions, and would be banned in many organizations.

All steels must be tempered following the quenching or hardening operation. The process of hardening should convert all the austenite into alpha martensite, which is a very hard, brittle substance. By heating to approximately 200°C the alpha martensite is converted into beta martensite, which has a greater ductility with little or no reduction in hardness. This will reduce the possibility of brittle failure.

Components which require toughness are tempered at a temperature of between 200°C and approximately 600°C. Tempering above 600°C will result in the material being fully softened and this, while it might be necessary in a limited number of cases for maximum impact resistance and ductility, is unusual.

The failure investigator with a brittle failure will be able to identify metallurgically and by hardness testing whether the problem is the result of undertempering, when the component will be above specification hardness, or overtempering, where the component will be too soft and can be expected to fail in a ductile manner.

If the structure is coarse grained, acicular and at the correct hardness, this will be the result of excessive time at the correct temperature or the correct time at too high a temperature, and not faulty tempering.

Tempering is normally carried out in air furnaces with fan-assisted circulation to ensure even temperature. It is essential that all the components reach the temperature and are held for a minimum of 10 minutes per 25 mm of ruling section.

Most heat treaters will temper with the components in the furnace for approximately 2 hours.

Double tempering is occasionally carried out on sophisticated tool steels. This is to convert any retained austenite into martensite. Where alloy steels are used which are very sluggish in their reactions, it is possible at quenching for some austenite to be retained. This cannot happen with normal alloy steels, but steels with significant amounts of tungsten and chromium present are susceptible to retained austenite. This can result in dimensional movement in service as the austenite slowly transforms to martensite. This results in a volume change as well as an increase in hardness. Only components with very tight dimensional tolerance standards such as gauges would be susceptible to this problem.

Variations in heat treatment

Up to now the discussion has been on treating the component homogeneously either by annealing, normalizing or hardening and tempering. There are a number of other processs, many of which have duplicate or triplicate names for the same basic process and all of which are some sort of variation on one of the above.

For the record, a list of these variations or alternatives, with a brief description, in alphabetical order, are as follows.

Austempering	A form of delayed quench.
Austenizing	An alternative name for heating to above the upper critical temperature.
Bright annealing	Annealing in some form of controlled atmosphere.
Classical annealing	Loading components into a cold furnace, heating to above the upper critical temperature, and cooling in the furnace.
Controlled annealing	A name given to several variations of softening, for example controlled removal cold work, or for producing a specific magnetic effect.
Cyclic annealing	This is applied to high carbon, high alloyed steel in order to soften it. The components are taken above the upper critical temperature, cooled to 650°C, and this is then repeated to soften them. This results in a finer grain with well-distributed carbides, rather than a long term high temperature annealing.
Magnetic annealing	A specialist treatment for magnetic steels.
Marquenching	A form of delayed quench.
Pack annealing	Annealing in a box with cast iron or sand, to exclude air.
Post-heating	A non-specific term, which is most commonly applied to post-weld heat treatment.
Pre-heating	Self-explanatory – any heat applied prior to welding or other forms of treatment.
Quench tempering	Cooling for hardening in a warm liquid to reduce the full hardening effect. Not recommended for high quality work.
Refine	Normally applied to high quality carburized components. This is short-hand for 'refining' (hardening) the core after carburizing, before hardening the case.

Secondary hardening	A term given to the double tempering sometimes required on high alloy steel to convert any retained austenite after quenching into martensite. There can be storage at −20 to −140 °C between the tempering operations.
Self-annealing	A vague term sometimes applied to large components as an alternative name for normalizing.
Self-hardening	An alternative name for air quenching, only applicable to high carbon alloy steels.
Skin annealing	A surface softening process to remove cold work, normally applicable to certain age hardening alloys such as nickel chrome, but sometimes applied to high alloyed steel.
Softening	Alternative to annealing.
Spheroidal annealing	An alternative name to cyclic annealing.
Stabilizing annealing	A general name given to stress releasing operations to ensure dimensional stability.
Step annealing	A specific form of stress releasing where complicated shaped components are cooled in stages to ensure even cooling.
Sub-critical annealing	Softening of components by heating at 650 °C, that is, below the lower critical temperature.
Sub-zero treatment	Cooling to a low temperature, −40 to −60 °C to convert retained austenite to martensite.

A competent metallurgist will have the ability to identify from the mechanism of failure and his microscopic examination whether or not any process has gone wrong. There seems no point, therefore, in discussing in detail what in some cases are rather exotic forms of heat treatment.

There are, however, three separate techniques by which the surface of steel can be taken to a higher hardness, generally for abrasion resistance. These are:

Induction/flame hardening
Case hardening by adding carbon to the surface (diffusion)
Case hardening by adding nitrogen to the surface (diffusion)

A fourth technique is where the case consists of both carbon and nitrogen.

All of the above are frequently involved in failure investigation either because the process has been wrongly carried out or because the component was subsequently abused in service.

Each of the processes is briefly described, and the problems which can arise with any or all of them discussed.

Induction/flame hardening

For this process to be carried out it is necessary that the steel has adequate but not excessive hardenability. A carbon equivalent (CE) in the range of 0.6–0.8% maximum would be acceptable.

The process requires that the surface is heated to a temperature above the upper critical temperature, held there for sufficient time for the heat to penetrate the necessary amount by conduction, and then quenched. The quench can be extremely efficient as the cold steel below the heated surface will help to cool the material in addition to the quench applied to the surface.

Induction hardening makes use of induction heating, which can be controlled by the correct use of coils, dimensions, voltage, energy applied and time.

With flame hardening use is made of an oxyacetylene flame, which has a temperature somewhere between 3000 and 3800°C depending on the flame condition. It requires the skill of the flame hardener to judge by eye when the surface of the steel has reached the required temperature. This will be between 750 and 850°C for most steels.

With induction heating it is possible to carry out trials and to specify coil size, voltage, amperage and time. It is more difficult to control oxygen/acetylene ratios and pressures, distance of flame from the surface, etc. There have been more problems associated with flame hardening than with induction hardening. Badly controlled induction hardening, however, can do as much damage as flame hardening.

With high quality/integrity components, use will almost invariably be made of induction hardening, with test pieces produced under controlled conditions being used for metallurgical examination, and then to produce a detailed process and product specification. Test pieces will be examined as a matter of routine during production runs. While the same might be laid down for flame hardening, it is much more difficult to control the heating conditions in flame hardening than with the use of induction coils. However, components such as massive gear wheels over 1 metre in diameter with large teeth would be extremely expensive to set up for induction hardening.

The problems which arise with induction/flame hardening are firstly the obvious one of either excessive or insufficient heat. Excessive heat results in large grains and therefore a brittle structure, and could also have excessive case depth. Insufficient heat will give material which is underhardened and with too shallow a case depth. Both can give rise to failure, the second – that is insufficient heat – will normally cause less problems, for example distortion, while excessive heat can result in dramatic fracture from brittleness.

Where alloy steels are used, with a carbon equivalent above 0.8%, then a problem arises with hardenability. This means that the cooling effect of the mass of metal below the case gives excellent quench hardening. This can result in a sharp demarcation between the case and core. As discussed under carburizing, below it is essential that there is a correct merge, that is, that there is a range of hardness from the hard case above 700 DPN to the core which will seldom be above 350 DPN. If this occurs too rapidly as can happen with alloy steels being induction or flame hardened, then there will be a very sharp demarcation and shear forces can crack the interface.

There is a case history where a swivel pin failed by fatigue starting at the case hardening interface because of this problem. This problem is one which is not always appreciated, and can be overcome at production by induction or flame heating for a slightly longer time than is required for the desired case depth, with the component allowed to cool in air. This results in some initial case forming. The induction/flame hardening is then repeated for a shorter time and the component is then quenched. This technique allows a 'false' merge between the case and core, and overcomes to some extent the problem of sharp demarcation.

Case hardening by diffusion

Iron can combine with various other elements to produce compounds which are then capable of reacting to provide hardness. The obvious one is carbon, which forms iron carbide. Given the correct conditions iron will also form iron nitride (FeN). It is now known that iron will also form a compound with boron – iron boride. While there is a 'boronizing' process, it is exotic and the standard investigator is unlikely to be involved with it.

Carburizing

Carburizing is still carried out to a limited extent by packing the components with activated charcoal, but this is now quite rare. Carburizing will almost invariably be a gas/solid process (gas carburizing) which is complex but results in iron carbide forming on the surface. The process is carried out at approximately 900–920°C. At this temperature the iron carbide will diffuse into the core at approximately the same rate as it is produced. This gives the ideal merge where there is at the surface approximately 1% carbon present as iron carbide, which diffuses more or less evenly depending on the time and temperature. This results in a composite material of 0.8–1.0% carbon steel on the surface of a much lower carbon steel core (0.2–0.35% carbon).

An examination of the iron/carbon phase diagram (Figure 15.1) shows that to harden the core requires a temperature in the region of 850°C, while the case is hardened from approximately 725°C. High integrity, heavy duty components, will also be double quenched following the carburizing operation, first from approximately 850°C to refine and harden the core, then from 725°C to refine and harden the case, with the core being fully tempered. This is required to remove firstly the large grains produced by carburizing, then those in the case when hardening the core.

This quite expensive procedure is seldom carried out on commercial quality carburized parts. Modern steels contain elements such as vanadium and niobium which inhibit grain growth. The use of carbo-nitriding, discussed below, also eliminates the expensive double quench.

It is still quite common, however, for failure to be the result of large brittle grains, or too sharp a demarcation between the case and core.

Gas carburizing is carried out in retort type furnaces, sealed from the atmosphere, where the carburizing gases are produced from propane gas, injected in controlled amounts with air to give the active carburizing constituents of carbon monoxide, carbon dioxide, water and hydrogen. These react with the surface iron to form iron carbide at temperatures above aproximately 900°C.

Nitriding

Given the correct conditions, iron will form the compound iron nitride (Fe_4N) at a temperature of approximately 500°C. This is normally carried out in an atmosphere of ammonia, which has the formula NH_3. At 500°C this disassociates to give atomic nitrogen, which then combines with the surface iron. This is an inherently hard compound which does not require to be quenched.

This procedure is, however, much more expensive than carburizing, requiring specialized steels and a much longer time in the relatively expensive ammonia atmosphere. Therefore, it is not chosen as a process except when its properties are required, these being high hardness, and a compressive surface that improves fatigue strength. It also has the very significant advantage that it does not require quenching; thus distortion can be more or less eliminated. The component also retains its hardness at a higher temperature than with carburized case hardening. Short periods of heating at 350°C will not cause any dramatic softening.

'Tufftriding' is a proprietary salt bath process resulting in a nitrogen-rich case. It is sometimes applied as a very thin layer to prevent galling, but also as a wear resistant surface. It is occasionally referred to as 'the poor man's nitride'.

Cyanide hardening

Potassium cyanide (KCN) is a salt, the combination of potassium (K), carbon (C) and nitrogen (N). When this is heated it melts at about 500°C and can be used at temperatures up to 900°C. When a steel is immersed in this heated salt at about 500°C it will be found to form a case which is very similar to the nitriding case. If it is heated to 900°C it will form a case which is similar to the carburized case.

Any temperature between 500 and 900°C will produce a case which has carbon and nitrogen present. This cyanide hardening process was therefore found to be extremely useful in that relatively small components could be immersed in the molten salt at the correct hardening temperature for the steel involved. Thus a medium carbon steel of 0.4% carbon which hardens from approximately 800°C could be immersed in molten cyanide at 800°C for a time sufficient to form the necessary case and then quenched directly from the salt bath.

Cyanide hardening will seldom if ever be found today. One obvious reason is that cyanide salts are notorious for being deadly poison. While fatalities among operators were not excessive, although significant, it was the advent of controlled effluent and a better appreciation of the environment which ended the reign of cyanide hardening. Effluent control meant that the processing of the quench water used in cyanide hardening, and the disposal of the cyanide salts increased costs dramatically. The biological effluent control of modern effluent/sewage works could not work with cyanide, which kills off the essential bacteria. The cost of an effluent plant is much greater than the cost of installing a modern carbo-nitriding/nitro-carburizing plant.

Carbo-nitriding – Nitro-carburizing

This makes use of a gas furnace which has an atmosphere somewhere between that of a nitriding furnace and a gas carburising furnace. Exactly the same comments apply as with cyanide – the components are heated at the desired temperature, then quenched directly. This is achieved with no effluent problem.

Vacuum heat treatment

This is occasionally specified for heavy duty parts in expensive materials where surface finish must not be affected, and decarburization cannot be accepted.

The components are placed in a vacuum chamber, the vacuum produced, and then the components heated. This will take place very rapidly by radiation and the components reach the necessary temperature quickly. When, however, the heating source is removed, the cooling in vacuum will be extremely slow. Therefore, where quenching is required it is necessary to circulate a gas, very often hydrogen which is very efficient and can act almost as quickly as oil or water, resulting in a rapid quench.

The failure investigator, finding decarburization on a component which was specified as being vacuum heat treated, could indicate that the vacuum heat treatment had been faulty.

Investigating heat treatment problems

The failure investigator will encounter numerous examples of brittle fracture, fatigue initiation, and cracking resulting from local heating effects where the CE is above 0.5%. Stray electric arcs from careless welding or electric discharge, and local friction from grinding or abrasion can produce brittle martensite with cracking, which results in failure.

It must be appreciated that 700–900°C can be achieved locally with comparatively little energy, and that with the underlying steel at room temperature, a very efficient quench results. This is demonstrated by grinding cracks, where a thin layer of brittle martensite or cementite can be produced on the surface even when coolant is used.

All types of steel welding use heat sources well above the critical heat treatment change points. Oxyacetylene has a flame temperature in the 3000–3800°C range. This, however, is not concentrated on a local area, and the welder will normally require to have a significant mass of metal at, or close to, red heat before he achieves melting of the surface to be welded. This is at a temperature in the region of 1500°C. Thus this type of welding automatically has metal pre-heated, and when the flame is removed, cooling through the critical range of 700–900°C will be quite slow. It is thus unusual for gas welding to result in excessive hardness. It is not uncommon, however, for grain growth, and thus potential brittleness, in the heat affected zone.

With electric arc welding where the heat results from the ionization of the gas, temperatures in excess of 10 000°C are achieved over a very small surface. When the heat source is removed very rapid cooling can occur – that is quenching with a hard brittle structure. While some debate continues, there is a general acceptance that a hardness figure in the heat-affected zone (HAZ) above 300 DPN (Rc30) can cause problems.

This problem exists without embrittlement by hydrogen, but obviously any nascent (atomic) hydrogen will worsen the situation. This is controlled by pre-heating, in which the steel adjacent to the weld is heated to produce a 'heat sink', thus slowing the rate of cooling.

Excessive grain growth in the heat-affected zone can add to the problem, but can also cause brittleness without hardness. The problem of large grains has been discussed elsewhere, and there is no doubt that large grains, even with soft steel, can result in brittleness.

The control of interpass temperature would appear to be an attempt to control grain growth. Where multi-run welds are necessary there will be a pre-heat specified as a minimum temperature to prevent local hardening, but also the temperature between weld runs will be monitored as a maximum. This is designed to prevent excessive heat build-up, and thus grain growth in the heat-affected zone. Experience shows that at present the reason for specifying and controlling interpass temperature is not clearly understood, and the methods and techniques of measuring are many and varied. There is little evidence, however, that serious problems resulting in failure exist in this area.

Failures have occurred quite frequently where undercut, lack of penetration, and weld porosity exist as the prime cause of failure. It is possible that with no heat-affected zone grain growth or excessive hardness, brittle failure would not have occurred, although crack initiation and propagation might have taken place.

Brittle failure from 'stray arcing' or tack welds are quite common where the carbon equivalent is above 0.6%. This is where electric arc welding is used without pre-heating. Welders have been known to heat their electrode as a 'cigarette lighter', by striking an arc. Tack welds, where components are held in position prior to controlled welding, and tack welded lifting lugs result in very local intense heat which can cause cracking, which is not removed when the full weld is produced but acts as the stress raiser to initiate fatigue.

While without doubt the serious problems identified by electric arc high production welding no longer exist, there is evidence that lack of a clear understanding of the metallurgy of welding heat treatment is resulting in controls being applied which are not necessary, such as some sophisticated post-weld heat treatment.

Stress relieving can be defined as heating to remove mechanical stresses resulting from cold working, mechanical deformation or stresses caused by expansion or contraction. Stress relieving should never affect the metallurgy of the pieces. There is evidence, however, that some post-weld heat treatment is acting as a tempering operation, reducing the hardness of heat-affected zones to below 300 DPN.

There is now available the expertise to produce the technical details for weld procedures to ensure that any steel with a carbon equivalent of less than 1.0% can have high quality welds with heat-affected zone hardness values of less than 300 DPN. This would require control of pre-heating and inter-pass temperature, and with higher carbon equivalent steels the use of low hydrogen welding rods.

Faulty control at heat treatment operations is not unusual as a reason for failure. This can be because the hardening temperature has been too low, or the tempering temperature too high, resulting in the steel being soft with a low tensile strength. This will result in ductile fracture.

Where steels are hardened at too high a temperature, or held too long at the correct temperature, brittle failure can result.

The fact that many quite sophisticated components perform satisfactorily in service having had less than perfect heat treatment would seem to indicate that modern steels have considerable flexibility, and that the safety factor built into the design is more than adequate. In fact, in the author's opinion, there is a need to investigate why many industrial components enter service and appear to perform satisfactorily when it can be shown that the metallurgical condition of the steel is suspect.

The failure investigator should be able to identify quite readily the difference between a carburized case and a nitrided case. It can be more difficult to evaluate accurately the carbo-nitride or cyanide hardened case. This is generally much shallower than a carburized case and will have certain characteristics indicating the type of case involved. All case hardened components, including those of induction and flame hardening, have a hard surface which is abrasion resistant but also commonly very brittle.

A common problem identified at production but also at failure investigation is that of grinding cracks or grinding abuse. A grinding wheel when it is in less than perfect condition, or not correctly controlled, can generate considerable heat between the wheel and the surface being ground. This can result in tempering the case but can also to take the case locally above the upper critical temperature and then, because of cold steel below and the coolant used, rapidly quench a local area, resulting in what are called grinding cracks. These are always at right angles to the direction of grinding, are short (not more than 1–3 mm long), are parallel to each other, and are extremely characteristic for failure investigation. They can result in exfoliation of the case or fatigue propagation. Examples are known where the case has flaked off, resulting in serious abrasion and expensive failure.

The investigator can make use of a Nital etch on the ground surface. This shows the different constituents resulting from the tempering effect of the grinding wheel and also a white line of cementite associated with a grinding crack. Grinding cracks are only a problem when surface hardness is greater than 500 DPN and thus is associated with fully hardened or case hardened components.

It will be obvious that it is not possible under any circumstances to carry out welding on a case hardened component. It is, however, possible to weld an assembly and then to case harden. It will be appreciated that the case hardening of the weld might be completely different from that of the remainder of the component. It is a most unusual technique, but one which could contribute to failure.

Any weld spatter on to the surface of a case hardened component or any electric arcing, even of a very low nature, will result in local and highly dangerous hard spots being formed, almost invariably having minute cracking at the crater.

Failure of case hardened components is commonly the result of inadequate or excessive case depth. The purpose of case hardening for the majority of components is to produce an abrasion resistant surface which is supported by a relatively ductile core. The classic example is gear teeth, where two hardened surfaces are rubbing together under load. If these were not hardened, there would be pick up and wear from the load applied, as discussed under compression failures Chapter 7. The hardened case accepts the much higher loading without causing abrasion or cutting. If, however, the gear teeth are through-hardened, then a loaded tooth will not be able to bend and will crack at the root with failure initiating either from fatigue or more commonly by brittle fracture. It is now accepted that there must be a ratio of 3:1 for the core to case depth.

There is a history of splined gears where failure occurred because the bore is carburized in addition to the teeth, and the case on the spline and on the root of the teeth does not leave sufficient core to accept the loading. It is generally accepted figure that there must be a minimum of three times the core to case depth.

Over or under-carburizing can result in problems, with over-carburizing having the same problem as above, that is, insufficient ductility of the core. Under-carburizing will result in spalling of the case because it has insufficient depth, and as the core moves the case will crack and exfoliate.

It is possible with case hardening to produce a very sharp merge between case and core. This is generally the result of carburizing at too high a temperature. While this without doubt will produce a thicker case in a shorter time, it does not allow the case to diffuse as it builds up quicker than the carbon diffuses, giving the same effect as discussed under induction/flame hardening of a very sharp demarcation between case and core (p. 269).

The design engineer when specifying case hardening will almost invariably specify the type, that is, induction, flame, carburizing, carbo-nitriding (nitro-carburizing) or nitriding. The case depth and the core and case hardness will also be specified. Only in a limited number of cases, however, is the method of measuring the case depth stated, and this can vary significantly depending on the technique used.

There are three methods and two techniques of estimating case depth. The three methods all use the first technique of optical examination of cross-section.

1. To estimate what is normally termed 'the effective case'. That is, the case which is considered to be above 0.7% carbon. This will be relatively thin and it is unusual for it to be specified.
2. The metallurgist can estimate with some accuracy the case depth from the surface of the component to the end of the merge, that is, to where the core structure can be seen to start to change to the case structure. This obviously gives a completely different answer by as much as five times the first case depth measurement.
3. The most common method is to estimate the distance between the surface of the component and the mid-section of the merge. This requires some skill and experience on the part of the metallurgist but generally gives a sensible result somewhere midway between the first two methods, provided there is correct training and the techniques are standardized.

The second technique is to use hardness testing.

It is now common that case depth is specified to be a minimum hardness, a minimum distance from the surface, this to be measured by micro-hardness testing using a stated load. All of the above methods are chosen by different organizations/engineers/metallurgists.

The investigator, presented with a failure associated with case hardening, will commonly have the broken pieces submitted without information on the specified case depth or other technical requirements. The metallurgist will have the information from the visual examination, non-destructive testing if necessary, and possibly discussion all of which are added to by the metallurgical examination. He or she should then be able to form a reasonably firm opinion on the mechanism of, and possibly the reason for, failure.

A not uncommon problem is fatigue initiation resulting from a local hard spot in a critical area. This can be the result of faulty 'stop off'. It is possible to prevent carburizing by copper plating or painting with a copper compound. This prevents diffusion of the carbon ion into the surface of the steel. Any porosity in the coating will result in tiny hard spots because of the iron carbide at the surface. Nitriding is stopped off with either bronze or tin plating, or again with paint compounds and the same comments apply.

It is possible using sharp grit blasting to identify hard spots in what should be a soft area. The sharp grit will not cut the hardened surface but leave a shiny patch on a matt background. While this will find large hard spots, it is unlikely to find tiny ones, which can be the most dangerous. This is the reason why high integrity components such as gears for aircraft engines and similar applications are not stopped off during the case hardening process. These components are carburized or carbo-nitrided all over and are then machined to remove sufficient metal on areas which require to be soft prior to hardening. This is obviously much more expensive but is an indication of the serious manner in which these industries take quality control.

There is evidence that failures can occur for various reasons not connected with heat treatment, where the component is of a poor quality, as a result of excesssive grain size, over or under case depth, etc. This would appear to be further evidence that many commercial components are over designed. As has been indicated elsewhere in the book, this presents the problem to the investigator that emphasis has been given to a metallurgical feature of poor quality which has had no direct influence on the failure because of the over-large safety factor used.

NON-FERROUS HEAT TREATMENT

In alphabetical order the following metals form alloys which can be age hardened, that is, solution-treated and precipitation-hardened.

Aluminium	Copper alloy Magnesium–silicon alloys Zinc alloys
Copper	Beryllium alloys Chromium alloys
Nickel	Beryllium alloys Chrome–aluminium, etc. alloys
Iron/steel	18% chrome, 8% nickel Duplex – various types

All of the above can have their mechanical properties altered by heat treatment, this always being to solution treat, and then to precipitation harden (age harden), or to anneal.

The metallurgy involved is complex, but basically requires firstly that there are inter-metallic compounds formed. These can have complicated formulae with at least two metals in the molecule but are basically not different from compounds such as sodium chloride (NaCl), magnesium sulphate ($MgSO_4$) etc.

Secondly, the matrix metal must be capable of dissolving these inter-metallic compounds and retaining them in solution after rapid cooling.

Finally, the inter-metallic compounds must want to precipitate from the solid solution. This precipitation strains the lattice structure thus increasing the tensile strength. The reactions are sub-microscopic, that is, they are not identified in micro-specimens examined with the optical microscope. Thus it is not possible for the standard metallurgical investigator to report on whether the process has been carried out correctly or not.

With the advent of the electron microscope, the precipitation has been identified and explained. The effect of solution treating and ageing, however, were well understood before the electron microscope confirmed the reason. There are not the dramatic improvements in mechanical properties with this process which can be achieved with the heat treatment of steel.

Hardening non-ferrous materials

At present there is no thermal process which produces a case hardening effect on any of the materials involved. The three hardening processes used are solution treating, precipitation (age) hardening, and annealing. Each alloy will have specific temperatures for solution treatment and precipitation (age) hardening. The annealing temperature will almost always be in the same range as for solution treating.

The components are quenched from temperature for solution treating, and slowly cooled from the same temperature for annealing. Components annealed cannot be precipitation (age) hardened.

Solution treated components will generally be softer than those annealed, as some precipitation occurs with slow cooling. The fact that the normal metallurgical investigator cannot see the differences in the structure makes the visual examination, mechanical properties and chemical analysis more relevant than the micro-examination. This is discussed in Chapter 13 on Materials information.

The following test takes each alloy group and discusses the approximate temperatures at which the alloys should be heated and the problems which can occur.

Aluminium and its alloys

There are three types:

1. Copper
2. Magnesium–silicon
3. Zinc

There are numerous specifications within these, some of which are given with their uses in Chapter 13 on Materials information.

The solution treatment for these alloys will be performed at about 500°C. It is essential that accurate specification information is obtained as the melting or softening temperature is not much above that for the solution treatment. Each alloy will have a narrow temperature range for best results.

Solution treatment can be carried out in a salt bath which must be neutral, otherwise it will contaminate or attack the aluminium, or more generally in air circulating furnaces. It is essential that the quench is carried out as rapidly as possible without any delay in order to achieve full solution of the inter-metallics. Any inter-metallics which are not taken into solution or escape from solution because of faulty quenching will result in less than perfect solution treatment and this affects the subsequent ageing.

Correct solution treatment leaves the material in the softest possible condition, which can be used for forming but not for machining. Soft aluminium and its alloys are very difficult to machine as they tear rather than cut.

Where any of these materials require to be annealed for any reason, such as electrical conductivity, then the same temperature would be used but they would be slowly cooled in air. Where maximum softness is required they will be solution treated.

Where ageing/precipitation treatment takes place at room or low temperature, if any forming is to be carried out, it will be necessary to hold the components in a refrigerator at or below 0°C in order to prevent precipitation. This applies in particular to rivets, where the sequence of events is to solution treat, hold in the softest possible condition, fit and form the rivet head which then ages at room temperature.

The precipitation (ageing) temperature will differ for different alloys, some being as low as room temperature, the maximum being in the region of 150°C.

Problems which arise with aluminium alloy solution treatment and ageing are firstly that because of the closeness of the solution treatment temperature and the melting or softening point, serious distortion can occur. Also, care must be taken that contamination does not occur with solution treatment.

The failure investigator examining a mechanical failure such as fatigue or tensile fracture could identify roughly whether or not the material was in the correct condition by hardness testing. This does not, however, supply the same degree of accuracy as with steel, as there is less of a relationship between ultimate tensile strength and hardness.

Tensile testing would be necessary for accurate results and if these were out of specification then chemical analysis would be essential to identify whether the correct alloy had been used in the wrongly heat treated condition, or the wrong alloy had been used. The ageing treatment is time/temperature dependent, that is, the components can be taken to an exact temperature for a very short period of time to achieve full ageing, or be taken to a lower temperature for a longer time to achieve the same result.

The ageing curve increases the tensile strength with time/temperature to a maximum, and the tensile strength then decreases. It is thus essential for the maximum strength, that exact specification information is obtained and followed accurately.

It is good practice to choose a long ageing operation which will allow some laxity regarding slight under-ageing or slight over-ageing without disastrous results. If a higher temperature is chosen with little difference in time, either under-ageing or over-ageing can result.

Typical mechanical properties for the alloys involved are as follows:

Aluminium–copper alloys, UTS 360–450 N/mm^2, 0.1% proof strength 130–350 N/mm^2.
Aluminium–magnesium–silicon alloys, UTS 140–360 N/mm^2, 0.1% proof strength 150–240 N/mm^2.
Aluminium–zinc alloys, UTS 470–600 N/mm^2, 0.1% proof strength 270–460 N/mm^2.

Copper–beryllium alloys

There is some information that copper–chromium and some of the aluminium bronzes can have ageing characteristics. This is less dramatic and further information should be obtained if these alloys are involved.

Copper–beryllium can be the strongest non-ferrous alloy. An ultimate tensile strength of 1540 N/mm^2 and 0.1% proof strength over 1000 N/mm^2 in the fully aged condition are possible. This generally requires some cold work between the solution treatment and ageing.

The solution treatment for copper–beryllium is performed at about 750°C, with ageing temperature ranging from 315 to 470°C. The solution treatment condition for copper–beryllium is soft and ductile, and can be readily formed. After ageing its ductility is very much less and with a tensile strength in the region of 1000 N/mm^2, considerable force is obviously required to produce any deformation.

One point to realize with copper–beryllium alloy is that its electrical conductivity varies from the solution treated or annealed condition to the fully aged condition, where it is reduced. Further information on this is given in Chapter 13 on Materials information.

The failure investigator might be involved with mechanical failure where it would be necessary to identify whether or not the material was in the correct heat treated condition. Hardness tests can give a reasonably accurate indication but for critical results chemical analysis and heat treatment tests might be required along with tensile testing.

Where copper–chrome and copper–aluminium bronze alloys are involved, the increase in tensile properties is much less dramatic. The failure investigator would tend to use mechanical testing where a tensile type failure was involved. It would be advisable to ensure that specification information was available for comparison.

Nickel alloys

There are three: nickel–beryllium, which has very limited use, some Monel nickel–copper alloys, and nickel–chrome, which is by far the biggest usage of nickel age-hardening alloys.

The nickel–beryllium alloys are analagous in all ways with the copper–beryllium alloys, but do not have the same dramatic increase in tensile strength. Their use is limited.

Again the electrical conductivity of these materials varies with the state of their ageing. That is, an over-aged nickel–beryllium alloy will have better electrical conductivity than a fully aged one. One use for these materials is in electrical controls which operate in the 200–450°C range.

The failure investigator involved with these alloys can have considerable difficulty as very slight differences in temperature can result in different characteristics, and specialist advice may often be required.

There are a few nickel–copper alloys – such as K Monel – where additives of titanium and aluminium give some precipitation after solution treatment. This gives some increase in tensile strength, but it is not dramatic.

Nickel–chromium alloys

These will invariably have other elements present such as aluminium, cobalt, iron, molybdenum and titanium.

When these alloys were developed, they enabled the engineer to make the gas turbine successful. While there is some slight increase in tensile strength, it is the effect on the creep properties which makes these alloys of considerable significance to industry.

The alloys are solution treated at a temperature of 1060–1150°C depending on the alloy. In some cases they are quenched, in other cases they are air cooled, depending on the material involved. This is followed by an ageing operation at approximately 750–900°C, again depending on the alloy. Very often the components are part aged to improve machinability and finally aged following a 'skin' annealing operation.

All of these materials are very susceptible to cold working, and this can make the surface brittle. It was found that a process termed 'skin annealing', which is a very rapid heating to a temperature of about 1060°C followed by quenching, would remove the cold work without significantly affecting either the solution treatment or any ageing operation. The full ageing at 750°C takes considerable time, up to 7 hours or more, as these reactions are extremely sluggish.

Problems identified with these materials will normally be corrosion, tensile failure or creep failure. The failure investigator then has the problem of identifying, if possible, whether the material has been over-aged or under-aged and whether it was the initial heat treatment operations which were faulty resulting in poor properties, or whether the material has been abused in service. This can be extremely difficult, requiring considerable skill and experience on the part of the investigator.

Steel

Austenitic stainless steels have low yield (proof) strength, and in the annealed condition an ultimate tensile strength slightly above that of mild steel. Considerable efforts were therefore made to increase the tensile properties and some small success was achieved with a limited number of complex austenitic steels containing inter-metallic compounds which obeyed the rules for solution treatment and ageing. The control at heat treatment was critical and the increase in tensile strength was not dramatic; thus these steels found little employment.

Recent developments, however, based on the 17/4–17/8 chrome–nickel analysis have shown more success. There are now a series of these steels known under the generic name of duplex stainless steel, and covering a relatively wide range of analysis. Details of their specifications and their limitations are given in Chapter 13 on Materials information.

They require to be solution-treated with the temperature involved being in the 1000–1200°C range. All the steels will then be in the fully softened condition and most will be difficult to machine to achieve a reasonable surface finish. It is thus common that they are supplied solution treated and part aged, when satisfactory machinability is achieved.

As with all these materials the precipitation (age hardening) is a time/temperature effect; thus the time and temperature to give the correct mechanical properties at the final age should be supplied with the raw material.

At the time of writing there is no evidence that problems – that is failures – are being experienced or investigated in this area, but there is evidence that some components are in service with less than perfect heat treatment.

Under-ageing will result in components with lower than specified tensile properties, but almost always better ductility and generally adequate impact properties. Over-ageing, however, can in theory at least result in the formation of the dreaded sigma phase. This is the precipitation of complex chromium compounds dramatically reducing the impact properties, with some effect on ductility. The sigma phase precipitates out in the 400–650°C temperature range.

Titanium

Titanium has a phase change at 880°C, where alpha titanium changes to beta. This is similar in some ways to the ferrite to austenite (alpha to gamma) change which makes iron/steel the leader in metals.

To date there are no dramatic improvements in mechanical properties with titanium alloys, but some alloys show promise. These contain aluminium, molybdenum, vanadium, tin, etc. The use of the alloys at present is confined to the aircraft engine industry, and to a lesser extent the petrochemical field. They require sophisticated control at all stages from casting, forging, machining, and heat treatment.

At the time of writing the run-of-the-mill failure investigator is seldom involved in failure analysis of titanium products.

Case histories

15

This final chapter is a series of actual case histories. It is not possible nor desirable to reproduce in a book of this type failure reports giving names with specific evidence, or information which could be readily identifiable. The information has therefore been modified slightly as necessary to ensure it is not possible for anyone to directly identify a problem.

With the exception of the final small group of historical failures, the author has been involved with all the other case histories.

In some instances a number of investigations of a similar nature are brought together to give composite information.

The purpose of the case histories is to show that the approach given in the body of the book is invariably used to a greater or lesser extent during failure investigation.

Wherever possible the relevance of specific features of investigating techniques has been highlighted.

The case histories are grouped according to the six 'mechanisms of failure', that is, tensile failure, compression, shear (torque), corrosion, erosion and heat, with a final small group of historical failures. Chapter 7 of this book deals with the six mechanisms, but it was considered useful to give a brief summary here before each group of case histories.

GROUP 1 TENSILE FAILURES

These cover brittle failure, ductile failure, fatigue and bending. In all of these there has been a pulling action. The investigator, by visual inspection alone, will identify the tensile characteristic and may then require further work such as hardness tests and micro-examination to identify the reason for the failure.

Tensile failure cracking of teeth on a spiral gear

Investigation found this to be the result of grinding abuse.

This component was a spiral gear driven by a bronze pinion with the output driving a jack mechanism lifting hundreds of tons. Cracking was identified on some gear teeth during a routine inspection. It was confined to three of the teeth, all on the flanks of the teeth, and all the cracks were vertical. The cracks were parallel and extended over 75% of the circumference, indicating they were the result of grinding abuse. This was a visual defect.

The surfaces were cleaned and etched and it could then be seen that there were colour variations on the surface associated with the cracking. It could be seen that although the cracking was confined to 50% of the teeth, the etching effect was present on all the teeth. This etch is used to show the metallurgical condition, and in this instance had shown that the surface varied and was typical of 'grinding bounce', resulting in local hardening.

Micro-specimens cut remote from the problem showed no evidence of cracking and that the material was alloy steel in the correctly carburized and hardened condition.

No metallurgical problems could be found with the heat treatment but the micro-specimens at the cracking showed a white surface layer and this had initiated the cracking.

The evidence thus was that during the grinding operation the surface to a depth of 10–15 microns had been taken to a temperature above 750°C and then rapidly cooled, forming the metallurgical structure known as cementite, which is extremely brittle. This was the result of friction at the grinding operation with the cold metal below the surface acting as a very efficient quench. The investigator could not tell whether the very brittle layer had cracked at the grinding stage, or during operation.

The typical configuration of grinding cracks is that they are at right angles to the direction of grinding, are relatively short and are exactly parallel to each other. They present extremely severe stress raisers and thus fatigue propagation will often occur, either through the complete section or very often with exfoliation.

This is an example of the importance of visual examination, firstly seeing the minute cracks, then recognizing the significance of fine parallel cracks. A skilled investigator will not be surprised to find the thin surface layer of brittle cementite.

Fatigue (tensile) failure, spiral gears

Tiny stress corrosion cracks had initiated the fatigue. Discussion identified a non-scheduled tensile overload.

The equipment consisted of jacks used to raise a swingbridge over estuarian water. Failure occurred on one of the jacks by fatigue over 80% of the cross-section as identified at visual examination. Metallurgical examination found that stress-corrosion cracking less than 0.25 mm deep had acted as the stress raiser to initiate fatigue.

Other jacks were removed and magnetic crack inspection found one local area on each had well-defined cracks. These were examined and it was found that all the fatigues had initiated from tiny stress-corrosion cracks. These fatigue cracks had propagated across approximately 50–70% of the cross-section without fracture.

In discussion it was found that during testing there had been an incident when the jacks had been overloaded with this load retained for a significant time in the marine atmosphere. It was determined that the tensile load was much higher than specified, and that this and the corrosion had caused stress-corrosion cracking. Once this unusual load was removed, the stress corrosion did not propagate, but fatigue did from the nose of the stress corrosion-crack.

This is an example of the metallurgist finding the mechanism and the metallurgical reason for the failure, but identifying the cause of the problem only through discussion with others.

If the overload application had not been found it is quite possible that the stress-corrosion/fatigue information would have resulted in an expensive redesign to cope with the stress corrosion.

Stress-corrosion cracking of austenitic stainless steel jackets

Cracking was found to be chloride contamination from the insulation, stress imposed at the steam inlet flange fitting, or contamination of the steam.

It is now common practice to use type 316 austenitic stainless steel steam heated tanks in the food and drink industry. The tank heating is with a jacket approximately 5–7 mm wide which is held in position with dimple welds. There have been a number of failures of this type of jacket, associated with the dimple welding.

Cracking of a random nature was found to be stress corrosion from contaminated insulation. This was from the outer surface of the steam jacket, and was cured by using uncontaminated insulation.

Other problems found were always associated with the surfaces close to the steam inlet or outlet flanges. A number of these were examined and all the problems were found to be the result of stress corrosion resulting in steam leakage. Investigation showed that this was caused by the fitters pulling the flanges together, thus applying a tensile load to the steam jacket. This was cured by ensuring that all such flanges were in compression before being bolted up.

A number of tanks used in a particular location had steam leakage which was not explained by the above. Evidence of stress-corrosion cracking from inside the dimple was identified by micro-examination and various discussions took place with the users, the water treatment personnel and others involved. It was then identified that the valves used to control the steam were of the paddle type. This results in a rapid surge of steam when the temperature sensor requests heat.

Further discussion identified that the location of the equipment was close to the boiler supplying the steam. The water treatment personnel then identified that there was evidence of 'carry over' of the water treatment system from the boiler to the steam side. It was pointed out that water treatment would use oxygen scavengers to remove oxygen from the water, and that these chemicals coming into contact with stainless steel could prevent or reduce the oxide essential to the stainless properties of austenitic stainless steel.

The first corrective step replaced the paddle-type steam valves with ball valves, which allowed a more gentle steam entry. Secondly, a trap was fitted between the boiler and each of the austenitic stainless steel tanks requiring steam heating. Finally, it was agreed that once these modifications had been completed, the steam jacket would be treated with an oxidizing fluid, and then heated to 250–300°C to produce an oxide layer.

These investigations are excellent examples of what can be achieved when there is cooperation from all concerned. It is not unusual that the investigator is supplied with negative information which would remove any blame from the client.

Fatigue failure of bits used for rock drilling

This failure was of two drill bits with three legs, with hardened rotating bits. Failure was on one leg and resulted in considerable sums of money being spent on recovering the drill string and the failed bit.

Investigation showed the failure was fatigue, initiating on the inside of the leg, propagating approximately one-third of the cross-section, and then – a most unusual occurrence – fracture by rapid tensile failure from the outside to the nose of the fatigue. This was identified at the visual examination. By far the most common, almost invariable, propagation is brittle fracture from the nose of the fatigue to the other side of the component.

Macro-examination showed that there was decarburization on the inside of the leg. Discussion with the manufacturers and those involved in drilling could not clarify the situation as the drill was designed to operate with the tensile stresses loading the outer sides of the three legs. Thus the area where fatigue commenced should have been in compression.

Further discussion elicited that in this instance the hole being drilled was relatively shallow and the normal pressure of drilling mud was not being applied. The drilling mud pressure would normally push the leg inwards, putting a tensile load on the outer surface.

When the drill is operating normally, this pressure is sufficient to retain the outer leg in tension, but in this instance there was insufficient pressure and the vibration of the leg during drilling allowed the inner surface of one leg to become tensile loaded. Once the fatigue had propagated approximately one-third of the cross-section, which is less than might be expected for a component of this type, failure occurred from the tension side, that is the outside, to the nose of the fatigue.

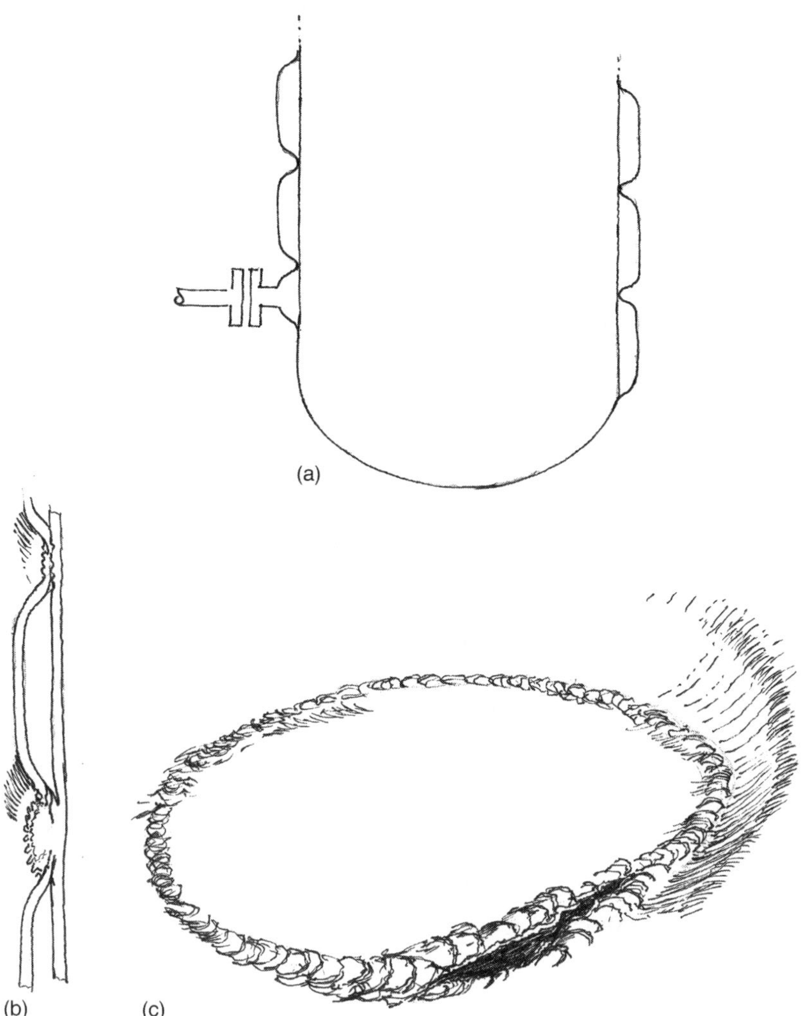

Figure 15.1 (a) The two flanges ready for bolting. When these come together, there is a tensile load applied to the outer jacket. (b) The steam jacket: approximately 5–7 mm width between the tank wall and the inner surface of the jacket. Dimple welds hold the jacket in position. (c) The local cracking on the dimple weld.

This information was reported and it is understood that the mud pressure was increased immediately, and no further problems occurred.

This is an example where the investigator had the ability and confidence to 'see the impossible', and then had the opportunity to discuss with operators possible reasons for it. Even so, there were still a few sceptics who believed there is such a mechanism as compression fatigue!

Overload fatigue failure of a ship propellor shaft

This was failure of a propellor shaft on a trawler operating in the North Sea.

The shaft was manufactured in austenitic stainless steel. Failure was by fatigue from the corner of the shaft keyway. Fatigue cracking had propagated with indications of torque over 70% of the cross-section before fracture occurred.

Metallurgical examination found no defect associated with the initiation. The keyway corner did not constitute a serious stress raiser and therefore the situation was discussed with those involved.

It was found that the harbour dried out at low tide, and that the skipper was known for his enthusiasm to be first out of the harbour on the rising tide, and last in on a falling tide. It was known that under some circumstances the trawler 'ploughed' its way out of harbour and 'ploughed' its way back in. The skipper was known as the 'fishing farmer'.

It was on entering the harbour under these circumstances that failure occurred. The results of the shaft fracture were quite serious in that as the trawler approached the quay and went into reverse, the propellor fell off. This meant that the trawler had no 'brakes' and struck the quay with some force, doing considerable damage.

This is an example of true overload where the failure investigator identified that there was no material defect or stress raiser and thus a design problem could exist. It was the intelligent discussion with outsiders that identified the overload cause.

Hot tensile stretching of gas turbine blades

A gas turbine engine had been removed from service because of turbine blade rub and a number of blades were submitted for metallurgical examination.

Gas turbines have the hot gases driving the turbine blades, which in turn drive the compressor discs. These turbine blades operate at temperatures in excess of 700°C and are subjected to high centrifugal loads. There is a warning given when the shroud at the outer end of a turbine blade touches the casing and this indicates that the blade has stretched more than normal and is reported as 'rub'.

This could be the result of a permanent stretch of the turbine blade because of creep but also the result of over-speeding, where higher centrifugal tensile loads are applied, thus stretching the blade within its yield

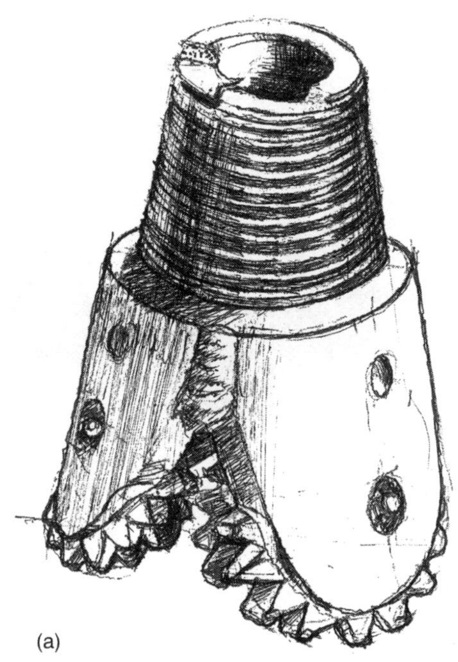

(a)

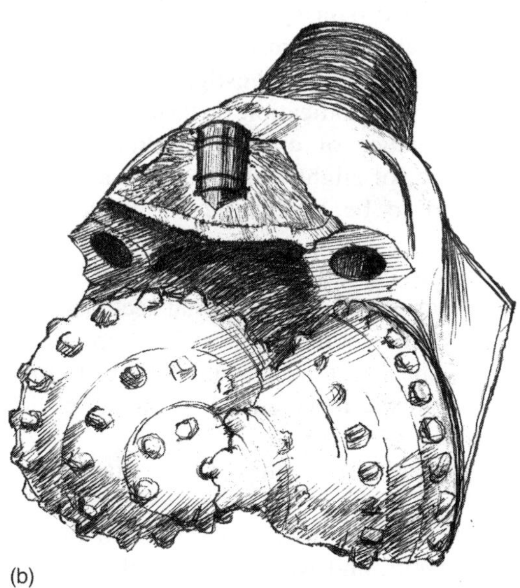

(b)

Figure 15.2 (a) A typical drill bit. (b) The failed bit, with one leg missing.

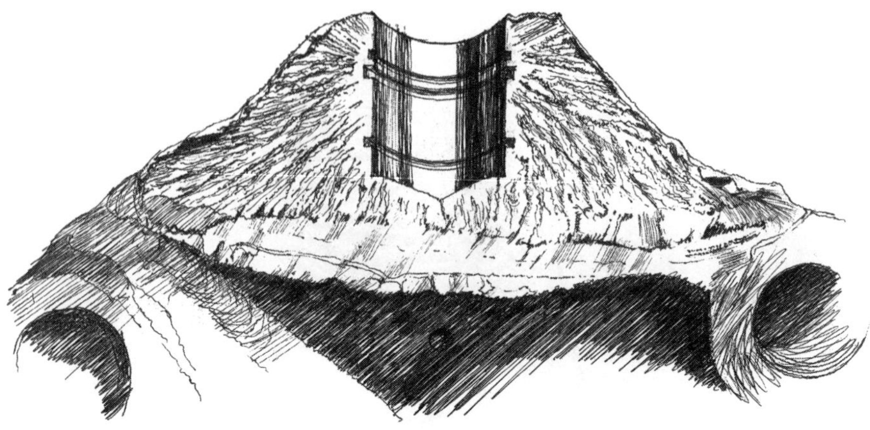

Figure 15.3 The fracture face, showing fatigue from the inner surface, then brittle tensile failure from the centre hole to the fatigue interface.

strength. The blade returns to its original size when rotation and over-speeding cease.

Examination of the turbine blades showed no evidence of creep and thus the engine had experienced over-speed and was not subject to creep. This meant that the engine could be returned for further service with replacement of the turbine blades used for the investigation.

Visual inspection showed indication of heat discoloration, but no excessive oxide and no evidence of cracking. Dimensional inspection found the blades to be close to, or slightly above, the maximum specified length. They were considered to be acceptable for blades which had been in service.

Micro-examination was carried out on a percentage of the blades to represent the total. This was on cross- and longitudinal sections, for which the polish and etch tecnhique for detecting creep voids was used. This allows the metallurgist to identify the minute voids which occur at grain boundaries when creep exists. It requires experience and skill but can positively confirm that no creep has occurred.

The blades which had not been cut for examination were returned to service. There are over 70 blades fitted to each turbine disc, and each blade costs a significant sum, thus the saving to the gas turbine engine operator was considerable, and was realized only because the metallurgist had the necessary skill and competence. The cost of a turbine disc failure can be catastrophic if the wrong information is supplied.

Brittle tensile failures resulting from faulty heat treatment

Failure is quite common with splined gears where the teeth are carburized for abrasion resistance and strength. The requirements generally are that the teeth are carburized only to a specified depth, but very often, because of lack of information or knowledge, the bore of the gear also is carburized. Thus where the root of a tooth coincides with the spline, carburizing of relatively small gears can result in insufficient core between the root of one tooth and the spline.

It is now accepted that there must be a ratio of at least 1:3 for case and core. That is, between the root of a case and either a surface or a second case there must be at least two-thirds the thickness of the core relative to the depth of the case. For example, a bar type component with a diameter of 25 mm should not have a case depth specified or applied significantly above 5 mm deep. This will then have a 5 mm maximum case and a 15 mm core. A 10 mm case would not be acceptable as this would result in a 5 mm core.

The purpose of the case is to give a hard, abrasion resistant, high tensile surface, with a core which will ‘breathe’ to allow some movement or flexing of the loaded surface. If this does not exist then the component is in effect through hardened and can be expected to be brittle.

There is a considerable history of splined gears, particularly relatively small gears, where the core between the root of one tooth and the spline is insufficient.

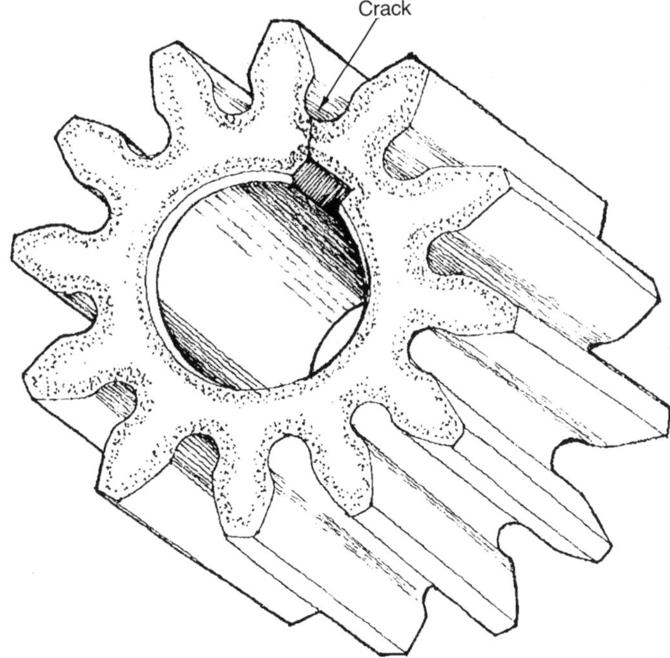

Figure 15.4 Section of splined gear showing the crack at the spline/tooth root.

This can result in brittle failure by the tooth being loaded and fracturing either from the root of the tooth or in the case of some driven gears from the root of the spline towards the root of the tooth.

The problem is generally associated with lack of communication or poor information. Where the designer specifies the teeth only to be carburized, giving the correct case depth, the heat treater, in a generous mood, may carburize all over, resulting in the core and the root of the spline also being carburized. In a few instances, the design has in fact specified carburising all over, this being a design failure.

Brittle tensile failures resulting from faulty heat treatment

This is a relatively common occurrence where the investigator can readily identify brittle tensile failure. In a few cases there may be a portion of the fracture which initiated by fatigue but this will be very small, with the major portion being brittle tensile failure. The failures are commonly associated with alloy steels, but plain carbon carburized components can also be involved.

The fatigue initiation could be from several causes such as over-tightening, stress-corrosion cracking, corrosion pitting, the presence of a sharp radius, grinding cracks, etc.

When the main reason for the problem is faulty heat treatment this can be of two types. Firstly, there are components which are correctly hardened but not correctly tempered. This results in a high tensile material with high hardness and low impact resistance and ductility.

Secondly, there are components which have been hardened from either too high a temperature, or the correct temperature being held for too long. This results in the material being in the correct condition regarding hardness, that is tensile, but brittle because of the large grain size.

Designers are aware that they require specific properties regarding tensile strength, ductility, yield strength and impact resistance. The drawing, however, invariably indicates this requirement only as a hardness. While this can have some relationship to yield strength, it can have no relationship whatever to either ductility as elongation, or brittleness measured as impact strength. The heat treater is then forced to use the hardness figure as the only criterion for acceptability, and this can be achieved by several routes which will not always give adequate ductility or impact properties.

Where high integrity components are involved, it is thus essential that test pieces are heat treated along with the components. For small components, provided the material is homogeneous and the heat treater has the necessary competence and commitment, then test pieces of approximately the same size will give reasonably accurate results if the heat treatment is homogeneous.

Where high integrity is involved, particularly with larger components, then it is essential that the test piece is integral with the component, that is, the test piece must be part of the component at the heat treatment operation. It must represent the most critical dimension of the component, and should be removed following heat treatment, machined and tested for tensile and impact properties.

There is a case on record where a person of considerable integrity but lacking in technical ability carried out test piece heat teatment in a laboratory-type furnace, and once satisfied regarding the test piece conditions, carried out treatment of the components which bore no relationship whatever to the laboratory-type furnace.

Where tempering is carried out at too high a temperature, or where hardening is carried out at too low a temperature, then this will result in failure of a ductile nature. In general this is much less dangerous and less costly than the brittle failures described above.

The experienced investigator will see the brittle failure at visual examination. Micro-specimens will show structures such as coarse martensite, well-defined grain boundaries, large grain size, serious surface decarburization, etc. This information, along with hardness testing, will allow a meaningful report to be produced in the majority of such failures.

Ductile failure of 'high seat' on truck chassis

This component was the seat at the rear of a truck used for loading components such as bricks, timber, etc. The seat had a high elevation to allow the operator to load accurately.

The failure involved was that of a truck loaded with timber which had to pass under a bridge to a rail-yard for unloading. The truck passed under the bridge, unloaded but later was found to be damaged, and an insurance claim was made.

Considerable discussion took place which identified the bridge as a possible cause. Only after accurate measurements were carried out was it found that when the truck was loaded there was clearance between the rear of the seat and the underside of the bridge. When, however, the truck was unloaded the springs raised the body up and instead of a clearance of 50 mm there was interference of more than 30 mm with the bridge, resulting in failure of the seat.

This is another example of failure investigation identifying the mechanism of failure rapidly, but the reason only after discussion and further non-metallurgical investigation.

Stud failures involving fatigue as the mechanism, and stress corrosion or over-tightening as the prime cause

A number of bolts, studs and set screws have been investigated where failure was by fatigue from the root of a thread. Investigation involving visual and metallurgical examination showed that prior cracking had occurred of a minute nature, less than 0.1 mm, and that the crack had acted as the stress raiser from which fatigue propagated. The visual examination identified that it was always the thread adjacent to the last engaged thread that cracked.

Various failures have been investigated all of which found a minute crack in the thread root which propagated by fatigue.

Some failures were caused by over-tightening. The calculation for converting torque to tensile stress is quite complex, requiring information on the loaded area, the number of threads, the thread pitch and the coefficient of friction. There is a dramatic difference between a stud and nut which are fitted without lubrication and a stud and a nut fitted with lubrication. The lubrication will reduce the friction by a factor of as much as 5, thus increasing the tensile load applied for the same torque reading.

It is also not always appreciated that zinc or cadmium plating will also reduce the friction, thus again increasing the tensile load.

Poor surface finish or corrosion can increase the friction, and thus reduce the tensile load for the same torque.

In addition to straightforward bursting tensile cracks, there can be stress-corrosion cracks, which are unique tensile cracks which bifurcate, or there is side-wall cracking. These can be more difficult to identify, and there is some evidence that their incidence is increasing.

Brittle failure of anchor swivel

Failure of an anchor swivel on a semisubmersible oil rig resulted in loss of an anchor, considerable down time and significant expense.

The swivel was examined metallurgically and found to have a large grained brittle structure. Impact tests gave figures of 10–12 J at room temperature.

Trial heat treatments carried out showed that impact properties could be improved to over 100 J. The situation was discussed and it was agreed that an attempt would be made to reheat treat the complete swivel. This was carried out under controlled conditions and samples taken from the scrap swivel confirmed that the situation could be recovered by correct heat treatment.

The figures from the trial heat treatment were above 75 J. The difference between these and the figures for test piece would be accounted for by the mass of the large component but clearly indicated that successful figures could be obtained. It is normally accepted that 30 J minimum at the operating temperature will avoid brittle failure.

With this information, samples were cut from the other seven swivels on the oil rig. These micro-specimens were seen to have the same large grained structure as the failed swivel and steps were taken to reheat treat all the swivels as the opportunity arose.

This is an example of an examination carried out in cooperation with the client where the reason for the problem was identified, and was shown to be recoverable in an economical manner. The cost of a lost anchor and the down time involved in recovering the situation was considerable.

It also shows, however, the importance of having the correct heat treatment specified where important components are involved, and ensuring that this is carried out.

Articulated truck 'fifth wheel' failure

This failure occurred when the tractor proceeded round a left-hand corner and the trailer carried on into a loch! The investigation showed that the 'fifth wheel' was fixed to the tractor chassis by two L-shaped beams. Failure had occurred with the two beams leaving the tractor chassis. It could be seen that each beam was held in position with three fillet tack welds, the tack welds being one at each end and one at the centre of the L-beam.

Visual examination showed that four of these tack welds had been fractured for some time with evidence of corrosion and abrasion on the fracture faces. Two of the fillet welds had fresh fractures and these were such as would be expected to put the fillet weld into tension as the truck turned left, with the load thus being pushed towards the right. All the fractures were in the weld metal.

Metallurgical examination showed that the fillet welds were of an adequate standard with no evidence of brittleness in the weld or in the heat-affected zone.

It was apparent to the investigator that with the loads involved the area of weld was quite inadequate. It was considered that the three tack fillet welds were designed for location purposes only to ensure that the beams were correctly aligned and adjusted, and that a full length fillet weld should then have been applied to both sides of the beams.

It was pointed out that no fault could be attributed to the driver or the maintenance engineer as the fillet welds were on the inner surfaces of the L-shaped beams, thus hidden from sight by the 'fifth wheel'.

In this instance it was considered that the information supplied by the investigator should be passed on to the manufacturers of this truck and, as far as the investigator is aware, this action was taken.

Hydrogen embrittlement failures

This case history covers three examples of hydrogen embrittlement failure.

The drive shaft of a gear box had wear between the shaft and bush. The bush was replaced and the drive shaft, which was relatively high tensile steel, was ground undersize, then chromium plated and replaced. This was an urgent operation with little time between chromium plating, grinding and fitting.

Fracture of the drive shaft occurred, this fracture being at the undercut where the chromium plating had been applied. Visual examination showed that the undercut had a relatively sharp radius, and tensile failure had occurred from the root of this radius.

Micro-examination and hardness tests showed that the material was of a satisfactory standard with good quality heat treatment and fine grained steel with a hardness typical of what would be expected for this type of component.

The component had been in service for a considerable time without failure and in discussion with the client it was not considered necessary to carry out impact tests, but the investigator was certain that these would have been reasonable for the steel used. The conclusion, therefore, was that hydrogen embrittlement was the cause of this problem.

Other examples of hydrogen embrittlement which have been investigated are a 1% carbon, 1% chromium bearing which was machined undersize on the outside diameter during manufacture, was then chromium plated as a salvage operation, and fractured when it was gripped in a chuck to carry out finish grinding.

Another example is of self-tapping screws used to hold fencing in position on a bridge where zinc plating was specified. These failed by brittle action within minutes of being driven home, with the roadway being scattered with the heads of the self-tapping screws.

In neither of the above failures had the components been de-embrittled following plating. No metallurgical or visual effects could be found which could have caused failure, and in neither case was there evidence of an overload having been applied.

Other examples of hydrogen embrittlement are stray currents resulting in electrolysis, and electrolytic corrosion. It is possible but less likely that hydrogen embrittlement would result from galvanic corrosion. Certainly hydrogen would be produced at the interface but this would be in small amounts and only slowly.

It is now accepted in, for example, the aircraft industry, that components with a tensile strength greater than about 1000 N/mm^2 should not be subjected to nascent/atomic hydrogen. Components with a tensile strength of less than 600 N/mm^2 need not be de-embrittled as they have sufficient ductility to 'breathe', that is, to absorb the stresses from the hydrogen. Components with a tensile strength between these, must be de-embrittled following any plating or any other process involving atomic hydrogen.

Where it is necessary for some form of plating operation on high tensile components, these should be stress released prior to the plating operation and de-embrittled/stress released immediately following the plating operation. They should be handled with care during the operation.

Hydrogen embrittlement, as far as the author is concerned, is the result of nascent, that is atomic, hydrogen entering the interstices of the steel. Once there, they combine to form molecular hydrogen, which appears to have a higher volume than the same quantity of atomic hydrogen and thus exerts a pressure which is translated into an internal stress or tensile load. When, therefore, the additional load of torque is applied to the shaft, brittle failure occurs.

The cure for hydrogen embrittlement is either not to subject high tensile components to atomic/nascent hydrogen, or if this is essential, to make certain they are subjected to a de-embrittling/stress releasing operation immediately after the operation. They should then be crack tested.

All plating operations result in atomic hydrogen at the cathode. It is not always appreciated that hydrogen is a metal and therefore will be appear at the cathode during an electroplating operation.

Fatigue tensile-type failure resulting from poor radius on a loaded shaft

This case study involved sophisticated forestry equipment where a single operator sitting in the cabin of a JCB uses two clamps to grip a tree, saws the tree near the root, turns the tree at right angles and using knives fitted to the leading edge of the clamps, holds the tree with one clamp, loosens the other and moves this along the trunk, cutting off the branches. The trunk is then reclamped while the second unit removes the remaining branches. At pre-prescribed intervals the first clamp is withdrawn along the tree trunk and used to hold this while the saw cuts the de-branched trunk into lengths of an accuracy of plus or minus 1 mm.

Failure occurred by fatigue on several units where the rod which pushed the clamp failed. This presented a problem in that fatigue is a tensile failure, but the load applied was compressive. The component had no metallurgical problems, but it was noted that there was a very poor radius at the change of section.

During a site visit to see the unit in operation, the reason for failure could be observed. Once the end of the de-branching took place, the first clamp was retracted to its original position, where it could be seen that the cutter heads were in contact with the trunk and that the roots of the cut branches protruded giving a sideways movement to the drive shaft. This resulted in cyclic tensile forces causing fatigue.

This was reported and the unit was remanufactured with an adequate radius in area mentioned. No further problems were identified.

This is a further example of visual examination and discussion finding the cause of the problem rapidly and efficiently.

It is not always appreciated that fatigue is the result of a tensile load. It is quite impossible to have fatigue when a component is in compression and this is what puzzled the investigator when the process was described. It was only when the unit was seen operating that the reason for the fatigue was identified.

Ball-bearing with stress corrosion cracking on the track surface

This was a ball-bearing involved in sophisticated offshore equipment. It was well-lubricated and should not therefore have been subject to corrosion.

The equipment was condition monitored with increase in noise used as the control parameter. A noisy bearing was removed from service, and sent for investigation. This identified cracking on the outer ball race on the working surface. This cracking was not serious, with no visible evidence of cracking even at ×40 magnification.

Micro-examination, however, showed stress-corrosion cracking. Stress corrosion has unique characteristics in bifurcating or side-wall cracking, and this was clearly present on this bearing track. This was reported to the client, who checked the oil and found this had a low pH (acid), thus causing the problem.

The investigator pointed out that had the condition monitoring not identified this, within a short period the bearing surface would have 'plucked' where two cracks came together. Once this took place abrasive material would destroy the evidence of the original defect.

The majority of roller and ball bearings which break up and are subjected to investigation result in the answer being a lemon!

Occasionally defects such as carbide network are seen and occasionally corrosion can be suggested as the reason for failure when this is serious enough to give a visible effect.

This is an excellent example of successful condition monitoring using sophisticated equipment. Had this bearing failed there would have been expensive subsequent problems, and it is most unlikely that the reason for failure would have been identified.

It also illustrates how minute stress-corrosion cracking can be, in that the cracks were not found under normal visual examination.

Had the oil been changed before the report was issued, the cause of the corrosion would not have been found, and it is possible that manufacturing might have been blamed.

Fillet weld failure on oil storage tank

This investigation followed a dramatic occurrence where a tank used for storing diesel oil had been lifted on to the back of a truck. After approximately 100 km the flat bed of the truck became filled with diesel fuel and each time the truck turned a corner, some of this fuel spilled on to the wet roadway, resulting in over 200 accidents on a motorway, with trucks and buses losing control, turning through 180°, and generally causing chaos.

The investigator was informed that the location of the truck had been identified by police in a police car 'sniffing' and following the smell of diesel oil to the source, which was the yard of a transport company.

On the flat bed of the truck was a 2 cubic metre storage tank. The investigator could see that failure was of a fillet weld at the tank base which had allowed the diesel oil to escape.

Micro-examination showed that the fillet weld had an inadequate leg length, that is, the length of the fillet weld was too short. It also showed that when the tank was lifted with a significant quantity of oil in position, a load would be applied to the floor of the tank which in turn would load the fillet weld.

The tank was reputed to be almost half full when it was loaded, and when found was almost empty.

The investigator pointed out to the police that it would be impossible for the truck driver to identify the weld defect as it required ×10 magnification to see the crack. Also it took some time for the oil to seep out and to fill the flat bed, and it was only on the last quarter of the journey that spillage occurred. Thus the driver and transport company could not be blamed for the problem.

This is a rare example of smell being involved in a failure investigation!

Figure 15.5 Result of the oil spillage.

Tensile failure from a minute material defect in a finned copper tube
These tubes are used for heat transfer in an air conditioning system. They are manufactured from thick wall copper tube, which then has fins extruded from the outer diameter to increase the air contact surface. Inside the tube Freon is used as the refrigerant, with air passing over the outside to cool the Freon. Evidence of the gas was found in the air circulation and examination using pressure testing showed leakage on some tubes.

These were removed and examined visually, and no defects could be found. Unfortunately the tubes with leaks had been cut longitudinally and thus could not be pressure tested to find the location of the leak.

Dye penetrant examination gave no positive indication and it was necessary to examine all the surfaces visually. With the tubes thoroughly cleaned there could be seen some draw marks, but no evidence of material defects.

Micro-examination showed that the tubes had been manufactured by taking a relatively thick walled tube and forming the fins by an extrusion process, that is cold working. This resulted in scoring which was difficult to differentiate from cracking.

It was, however, found that on a limited number of fins there were crack-like indications at the roots of the fin. Micro-specimens through two of the suspect fins identified that the operation had resulted in thinning of the wall at the root of the fins. There was then evidence that this had resulted in the pressure from inside the tube causing fracture of this thin wall, resulting in the Freon escaping.

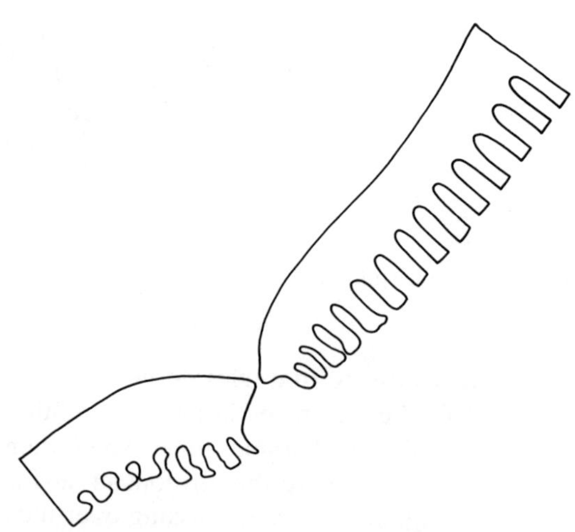

Figure 15.6 The finned tube with the reduction in wall thickness.

Examination of other areas on the tube confirmed that there were several lengths where the wall thickness was eccentric, that is, greater pressure had been applied to one portion of the circumference than the other, resulting in thinning of the wall at the root of the fins. This, therefore, confirmed that the defect was the result of manufacture. It was identified that manufacture had been carried out some considerable time ago in a foreign country; thus it was agreed that no further action could be taken.

This is an example of a badly conducted investigation where considerable additional expense was incurred because of lack of logical procedures which had made further investigation difficult. If identification of the leaking tubes had been isolated for use in a sensible manner, then these tubes could have been re-pressure tested either in complete lengths, or in short lengths if necessary, to identify exactly the location of the leakage. The tubes, however, were sent to the investigator after cutting and this considerably increased the cost of the investigation.

GROUP 2 COMPRESSION FAILURES

Following are case histories where compression was identified as the failure mechanism. This can result in distortion or bursting but seldom fracture. This is where pushing is involved, and it must be realized that up to the yield (proof) strength of materials, this factor is the same whether pulling or pushing occurs.

Once the yield point is reached, with tensile loading the cross-section is reduced, and thus the unit stress increases. The reverse occurs with compression loading, for which the unit stress to cause failure must be increased. Hence the fact that this mode of failure is quite rare.

Compression loading of a large nut, resulting in shear failure of the threads, releasing the riser on a drilling rig

This failure was of an extractor tool used to lift an offshore drilling riser. This is the metal tube, up to 0.75 m in diameter, inside which drilling takes place. It runs from the underside of the drilling rig to the seabed. When it is necessary to lift it, an extractor tool is screwed on to the top portion of the riser, which is then lifted from the seabed. During one such operation failure occurred of the threads inside the extractor tool, and the riser fell to the seabed.

Examination of the riser showed no problem, but some of the extractor tool threads were trapped in the male threads. These could be readily removed, leaving the riser undamaged. Failure could be seen to be by shear of the threads of the extractor tool.

Examination of this showed that it had 20 threads, of which 5–6 were engaged and had failed. Some of the failed threads were available for examination from the riser and others which had been collected when the tool was removed.

Examination of the remaining threads of the extractor tool showed these were corroded, had debris inside and had not been engaged for some time.

Further examination of the failure when the failed threads were placed in position showed that the underside of the threads had been subjected to compression with shear failure occurring by pushing the extractor upwards.

It was therefore reported that the extractor had not been fully engaged, and that it was probably some time since it had been fully engaged. It was indicated that the engagement was probably sufficient to lift the riser when there were no problems. Failure had occurred when, with the extractor fitted, the rig had risen and then fallen, the weight of the rig pushing on the threads. This is therefore an example of compression failure.

Hardness tests carried out on site and micro-examination of portions of the threads available showed that no problems existed with the steel of the riser or the extractor tool.

This example illustrates the importance of accurate visual examination, and an understanding of the working of the equipment.

Compression failure of a ratchet and pinion, where a foreign object overloaded the teeth

The most probable reason appeared to be sabotage.

Lorry-mounted cranes operate with a ratchet which is moved along its length by hydraulic cylinders, one operating in each direction. Engaged with the ratchet is a pinion which swivels the crane boom. In the central position the pinion is located at the mid-point of the ratchet.

Failure occurred when the boom was unable to move in either direction, and investigation showed impact damage on four of the teeth of the pinion, and six of the teeth on the ratchet. This damage on the ratchet was at about the mid-section.

The investigation found failure was the result of a foreign object embedded between the teeth of the pinion and the ratchet. It was suggested that this would appear to be a ball-bearing but no evidence of any foreign object was submitted with the component.

The conclusion indicated that the most likely reason for the failure was sabotage, as in discussion with those involved there was no other way which could have resulted in an object such as a ball-bearing becoming trapped in a closed system.

Examination showed that the teeth had been subjected to severe compression, resulting in failure of the case hardening on the pinion, and fracture of three teeth on the ratchet.

Micro-examination and hardness tests showed no material defects, with good quality case hardening, a satisfactory merge between case and core, and hardnesses which would be expected from components of this type. It

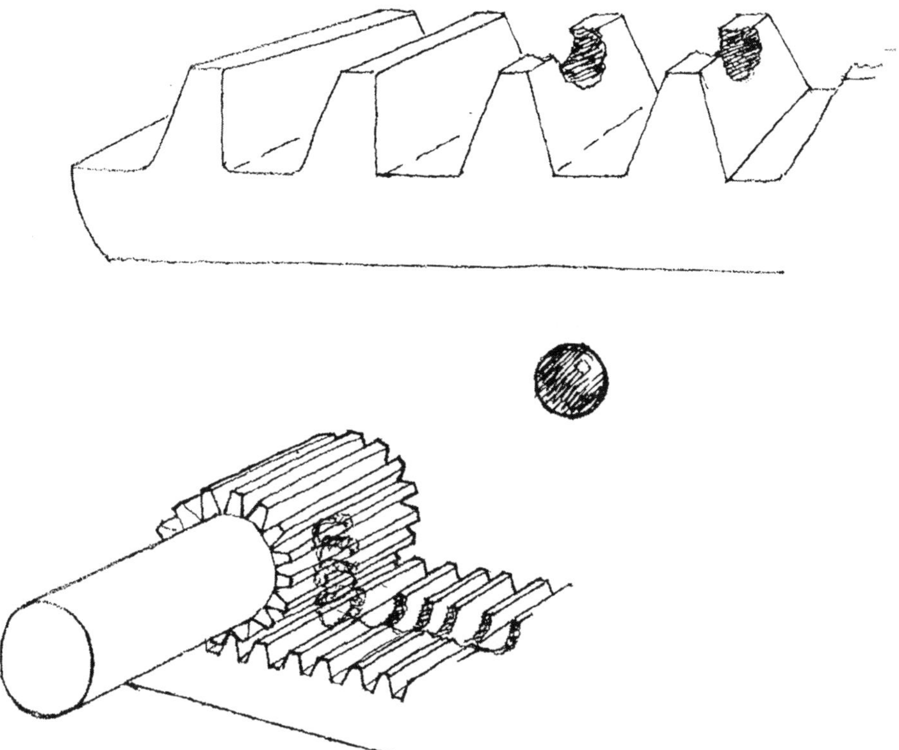

Figure 15.7 The local damage to the rack and pinion teeth. Whilst no ball-bearing was actually submitted, the evidence indicated that this was the source of the problem.

had been hoped that the micro-examination would have supplied some information regarding the type of material causing the compression, but no such information was available. This was reported as further indication that the damage was caused by an object which was at least as hard, if not harder, than the pinion and ratchet. This would suggest something like a ball-bearing.

It is not uncommon for failed pieces to be submitted for examination along with information from the client that suggests sabotage or deliberate damage. This failure, however, is a rare, almost unique example where there must be a high probability that a ball-bearing or similar item had been involved, with no actual evidence nor any indication where this had come from or how it could have accidentally entered the system.

Surface break-up of impacted tappet of rock-breaker

This failure was of tappets used in earth/rock-breaking equipment. The operation involves a cylinder to which impact is applied by air pressure, which in turn strikes the tappet, which in turn moves forward to impact on a chisel.

A specific batch of rock-breakers was found to have problems in that there was severe abrasion resulting in oversize of the cylinder and air leakage. The abrasive material was found to be from the break-up of the working surface of the tappet. Micro-examination showed that the tappet was manufactured in alloy steel. It was surface hardened but showed signs of micro-segregation. This resulted in ferrite and martensite striations, and it would appear that the reason for failure was that as the tappet was struck, the hardened portion (martensite) moved because the parallel striations of ferrite, which is relatively soft and ductile, lacked support. That is, instead of the whole surface being homogeneously hard and therefore moving as a platform, tiny areas moved in relation to the softer area. This resulted in particles being removed from the surface, acting as an abrasive, and thus causing the problems found.

It was recommended that in future the material should be specified to be free of micro-segregations, and if necessary trial batches should be subjected to a simple micro-examination.

Marine engine clutch plates and rubber mixer, both showing compression overload

The clutch plate was from a marine diesel engine with the clutch being a series of mild steel plates. These were located by teeth on the outer diameter and when running in the forward direction the shaft rotated within the bore of the clutch plates. When reverse was required a load was applied to the clutch plates and these were brought together to reverse the direction of rotation of the shaft.

A serious problem occurred, in that when reverse gear was selected, the propellor rotated in the reverse direction but could not then be disengaged. This occurred twice. In the first instance the clutch was replaced. In the second instance, however, the clutch plates were submitted for investigation.

Visual examination showed that the clutch plates had been subjected to severe compression resulting in extrusion, which reduced the bore diameter and increased the outer diameter. This eventually resulted in the clutch plates bore contacting the shaft and when the teeth were disengaged the shaft continued to rotate.

Micro-examination confirmed that the cold work was in effect causing extrusion.

It was suggested that reducing the spring pressure on the pack of plates would cure the problem.

The second failure was of a rubber mixer. This in essence is like a large dough mixer with a chamber inside with rotating vanes. When the rubber and additives are fed into the chamber, the rubber is cut and compressed, the vanes forcing it in a single direction and gradually increasing the load, thus mixing it with the other constituents. These constituents are oil, a filler material such as clay and sometimes a plastic extender.

Failure was identified in the first instance by the end wall of the mixing bowl cracking. Investigation showed that this had been displaced outwards, and further examination showed that the extremities of the mixing scroll had evidence of compression failure.

Further investigation and discussion with the operators found that this unit was used to its full capacity, whereas similar units operated at a much lower level of loading and compression. The electric motor involved had a rating of 4000 horsepower and it was only when this was used to the full that the problem existed.

The cure of this fault was firstly to strengthen the end faces of the mixer bowl, but also it was found necessary to leave greater clearance between the periphery of the rotating component and the inner diameter of the mixing bowl.

Micro-examination and hardness tests proved in both instances that the distorted material had evidence of severe cold work; hardness tests confirmed a significant increase in hardness.

GROUP 3 SHEAR/TORQUE FAILURES

These case histories relate to shear, which is the guillotine action, and also cover torque – that is, twisting.

Shear is in effect a combination of tensile stress and compression. The guillotine or punch pushes the top and applies a tensile load at the underside.

Pure shear failure is seldom encountered in failure investigations, but torque is a relatively common mechanism. A torque connotation can commonly be found with fatigue and tensile mechanism. This produces angular striations, at 45° to the fracture face.

Shear (torque) failure of studs, which had been grossly over-tightened
This case history was the investigation of a swivel-type bucket dredger barge, where in operation the bucket swings at right angles to the barge, drops rapidly into the water and then closes. The bucket is then lifted and swung through 180° to drop the debris into a hopper.

The equipment had been recently overhauled and after comparatively few operations the crane portion of the dredger toppled into the water. Fortunately the operator scrambled clear and no one was injured.

Investigation was requested and it was seen there were 32 setscrews retaining the inner track of the three-metre diameter roller bearing which swivelled the crane. All 32 studs had failed. Of the 32 setscrew failures, 12 were torque failures and all 12 had the head of the setscrew tack welded to the upper side of the bearing.

Of the remaining 20 setscrews, 15 showed evidence of fatigue initiating at the thread root, with the fatigue propagating in some cases by over 90% and in others less than 10% before fracture. There were also 5 setscrews which had failed by ductile tensile failure from overloading. Thus it could be seen that the dredger had been operating for a period with only 5 of the 32 setscrews intact, that is at approximately 15% of the design specification.

Micro-examination and hardness tests showed that all the studs were metallurgically satisfactory and had been produced by thread rolling. The micro-examination confirmed that the studs which had failed by torque were also satisfactory, and the visual examination showed that these were the initial failures, their having most corrosion on the fracture faces. The five setscrews which had failed by ductile tensile failure all had fresh fracture faces.

It was ultimately reported that this failure was the result of over-torque tightening 12 of the setscrews during the assembly operation. The operators must have been aware of the problem and had tack welded the heads to stop them falling out. This abortive corrective procedure had resulted in fatigue overload of other studs as the bucket was lifted. The fatigue variation could be explained by the fact that the first stud to fail by fatigue had the support of the other studs and that as the failures progressed, the tensile load to cause tensile failure increased. When the last fatigue failure occurred, the load remaining was sufficient to cause tensile failure of the five remaining setscrews.

This would indicate a reasonable safety factor, as out of 32 specified studs, 27 had fractured before failure – an approximately 85% safety factor.

During discussion it was identified that the assembly required the setscrews to be torque loaded to a known load in dry conditions. It would appear that this had not been carried out and there was visual evidence that grease had been applied to the setscrews. This would mean that the torque load which would have been adequate to tighten the setscrews in the dry condition was excessive when applied with lubrication. Once a number of setscrews had fractured in this way, the torque must have been reduced and the assembly completed.

It has been pointed out elsewhere in this book that there appears to be some misunderstanding regarding the relationship between torque, friction and the tensile load which is then applied.

This is an example of the investigator seeing the quite obvious failure mechanisms of torque, fatigue and ductile tensile failure on studs which were within the technical specification, where the less obvious evidence of grease associated with some studs gave the lead to the true cause of failure.

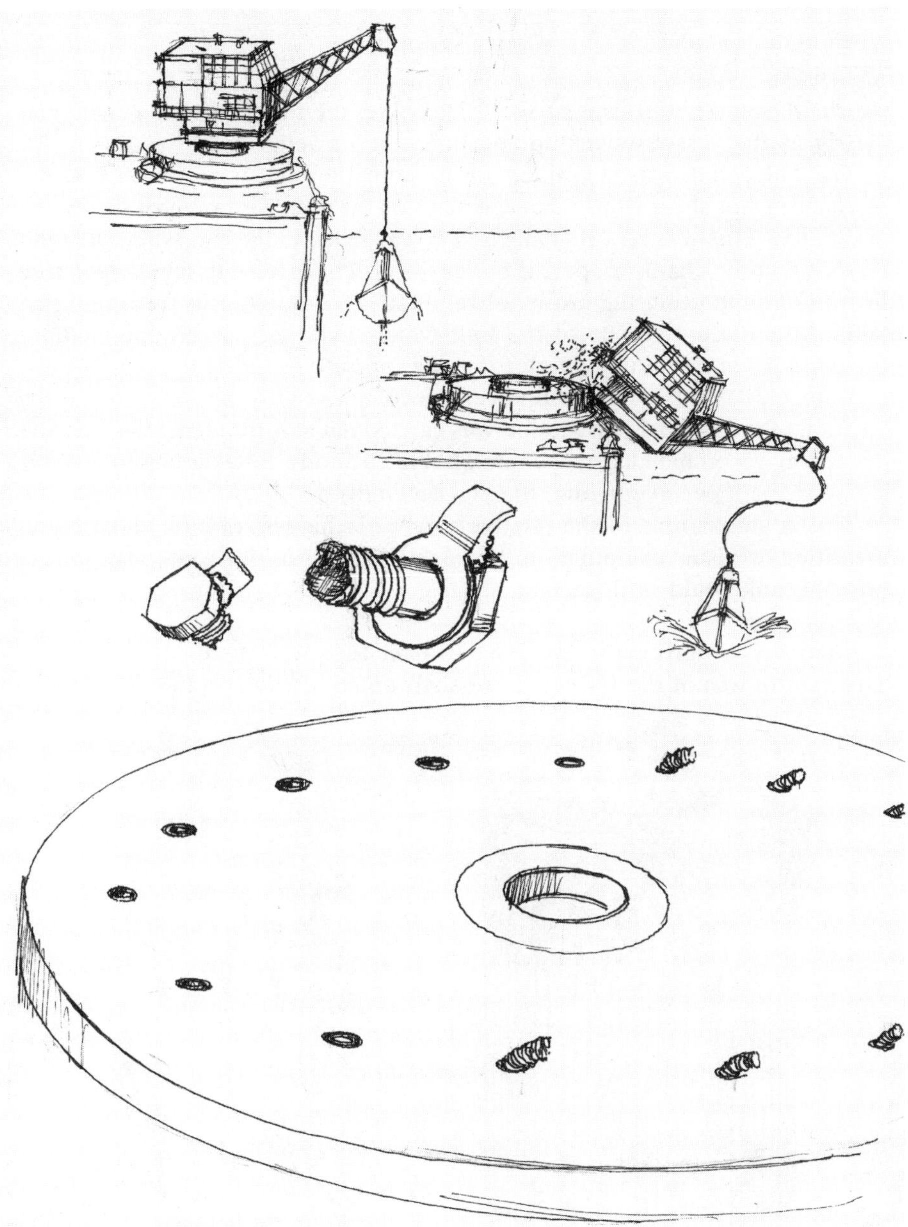

Figure 15.8 Failed bucket dredger with enlarged mounting stud illustrating the torque failure and the tack weld on the head.

A rare pure shear failure

This is an example of *guillotine* action on an anchor chain link, which was not correctly held in the chain stopper at the time of the incident.

The chain, which was used in offshore oil drilling was held in the winch with a stopper designed to fit snugly at the crown of the link and thus support the chain over a significant surface. This takes the full load of the chain and the rig as the vessel moves up and down in a swell.

The failed link was submitted for examination and it was seen that the mechanism was shear from a damage mark on the outer surface near the crown of the link. This was reported and a visit was made to the rig, where it could be seen that the chain stopper, if not correctly engaged, dug into the surface instead of supporting the crown of the link. This resulted in loading a line around the circumference of the link which acted as a guillotine. Failure occurred when a relatively high force was applied in gale force conditions.

With the layout of the equipment it was not readily possible to see the location of the stopper. Inspection was carried out at infrequent intervals and the report recommended that there should in future be examination of the chain and stopper every time the rig was moved.

This is an example of the relatively rare mechanism of pure shear being identified by visual examination. It was then necessary for a site visit to find how the link could be loaded in shear.

Splined drive shaft with overload impact torque failure

This failure was of the splined drive shaft on an offshore crane. This failed

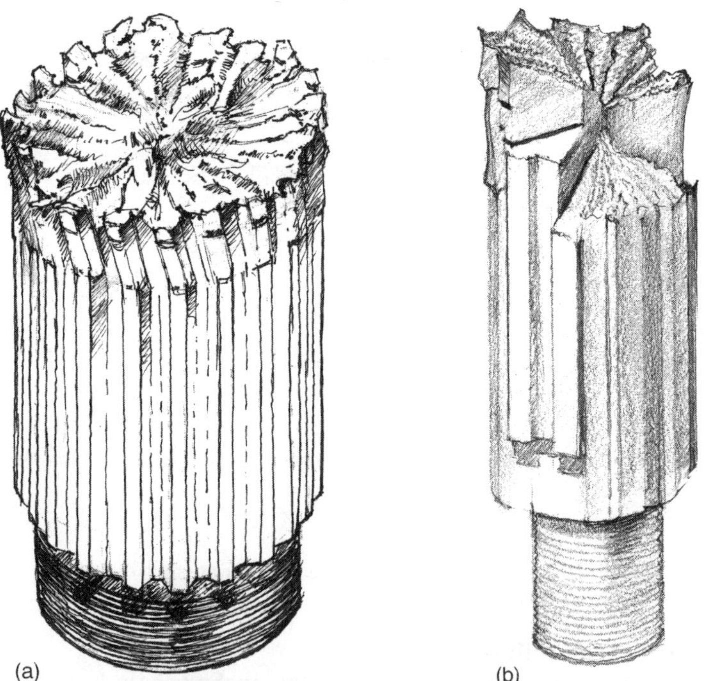

Figure 15.9 (a) The characteristic visual effect of impact torque where stress raisers such as splines are present. (b) Where the torque load has been applied in stages, resulting in the 45° angle fracture face.

and investigation showed that it was a rare example of impact torque. Failure was by rapid cracking from the root of the splines towards the centre.

This is an unusual type of failure and requires the very rapid application of a high torque load to a surface which has stress raisers. Metallurgical examination showed that the material was in the condition which might be expected for a component of this type and that it was a correctly heat treated alloy steel with a hardness giving the expected tensile properties.

The situation was discussed and it was identified that at the time of the failure a problem had arisen with the power supply to the crane. The crane operator was unaware of this and was manipulating the crane which had a heavy load on the supply vessel deck. The sea was rough and the supply vessel was heaving to an appreciable extent. With the operating lever applied in the lift condition, the power came back on and instantaneously the crane started winding rapidly inwards taking up the slack.

As the crane lifted, the supply vessel dropped lower in the water and the load was lifted only about half a metre before it fell back onto the supply vessel. This explained the impact torque.

An identical failure has been examined where the automatic stop on a rail-mounted crane failed. The crane travelling at full speed with a load was stopped by impacting the end wall.

Failure examinations to increase the shear/torque strength of a racing car drive shaft

This is the saga of a development project on halfshafts for a racing car in the days when racing cars had halfshafts. This car had an extremely efficient engine, an excellent driver and provided the car left the starting line, he could win races! The driver would sit at the starting line with both feet hard on the floor and when the flag went down, his left foot came up and if one of his halfshafts didn't fail he would frequently win the race.

Investigation showed that when failures occurred they were pure ductile torque failures. There was also evidence of less than perfect machining at the change of section where the torque failure occurred. This was close to the radius where the shaft section changed to the splined section.

The first development, therefore, was to improve the surface finish and use a stronger material. This was reasonably successful with the car winning a number of races without a change in procedure, that is both feet hard on the floor and left foot raised at the start.

Failure then occurred by fatigue initiating at the radius of the spline. The fatigue started at the end of the spline. It could also be seen that there was chatter present and that wear on the spline had taken place before cracking commenced.

Trials were therefore carried out with induction hardening in this area of the shaft. This was not successful, with failure occurring on the first start in a brittle manner where, instead of fatigue, there was brittle/torque failure.

This could be seen to be because of the sharp demarcation between the case and core, as there was a shear at this point which propagated rapidly to tensile torque failure.

The hardening was therefore changed from induction to carburizing, with the best quality carburizing steel used to give the core strength along with the necessary blend between case and core, and a good quality relatively thin case. This was initially successful but magnetic crack detection identified cracking of a torque nature, again in the area of the radius to the splined end. This was opened and found to be, as expected, fatigue. As this fatigue started on a perfect radius with no stress raiser present, it was realized that the fatigue strength of the steel had been reached.

With some doubt as to the outcome, case hardening was tried on the shaft itself and the radius to the splines, in addition to the splines. This resulted in brittle failure at the first start, again with torque. This was not unexpected as the shock load of the rapid start could be expected to crack the case, and with the high loads involved immediate fracture could be expected.

The final and most successful shaft was made from what is in effect a spring steel, 0.5% carbon, 1% chromium, 0.2% vanadium. This was hardened and tempered to give a tensile strength in the region of 1500 N/mm^2 (100 ton/in^2), which was then nitrided on the splined area only. This proved to be a success and many races were won with no problems with the shafts.

This is an example of a development over a relatively short period of time which could only be financed by a very rich man with a passion for racing. It also required excellent cooperation between the engineering designer, the metallurgist, and the heat treaters.

There had been no compromise whatsoever on behalf of the rich driver, who insisted on lifting his left foot as rapidly as possible as soon as the starting flag was lowered.

Shear cracking initiating fatigue

The shear crack was from the case/core interface on a heavy duty swivel pin on earth-moving equipment subjected to high loads as the tractor turned either to change direction or to reverse. Fracture of the pin had occurred with resultant serious and expensive damage.

The investigation showed that failure was by fatigue cracking, then tensile fracture.

Fatigue had initiated associated with the case hardened pin, but it could be seen that the fatigue was not from the surface but from the case/core interface.

Micro-examination showed this to be the result of too sharp a demarcation between the case and core. No other material defect could be found.

It was identified that this high quality alloy steel pin had been hardened by the induction process and it was thus recommended that in future such pins should be given a double induction heating, the first air cooled to produce a 'false case' followed by the second oil quenched to give the effective case.

The problem has been identified that alloy steel when induction hardened can undergo very rapid quenching from the outer surface and from the cold steel, resulting in a very sharp demarcation, causing cracking by shear at the interface, and the shear crack then propagating by fatigue. A satisfactory merge from case to core would have distributed the shear stresses.

This is an example firstly of the importance of detailed visual examination: this showed that the fatigue initiated below the surface, which is very rare. Secondly, the metallurgist could recognize and appreciate the significance of the sharp demarcation between the case and core.

Finally the report recommended the appropriate technique to obviate further failures.

Figure 15.10 The fracture face, with the fatigue initiating at the merge point, not the surface.

Shear/torque failure of truck drive shaft from overload

This investigation was into the drive shaft on a heavy duty vehicle which had failed during start-up. Failure was identified as being by torque applied in a gradual manner and the investigation found no evidence whatsoever of any metallurgical or material defect.

Discussion identified that this shaft was driven through a gearbox with a constantly engaged gear train where the gears were actuated by air pressure. In discussion it was found that if a vehicle of this type was left without being used for a period of more than 7/10 days, it was possible for the air receiver, which was fed by compressor, to empty. The driver was instructed under these circumstances to run the engine in neutral, thus allowing the pressure to build up.

In this instance it was a partially trained driver not aware of this instruction who engaged first gear. When nothing happened he increased the revolutions until at a certain point the compressor had sufficient air for the gear to engage.

This failure identified two things. Firstly, under these circumstances the brakes held and resulted in torque being applied to the drive shaft, which was therefore the weakest unit in the train between the crankshaft and the wheels. The procedure was amended and on every windscreen there was a clear instruction to drivers not to engage first gear at any time until the engine had been running for a certain minimum time.

Secondly, the system was redesigned with a fail-safe component which fractured at a lower level of stress than the drive shaft.

Visual examination showed that the drive shaft had a typical torque failure with the fracture being at 45° to the normal and the surfaces having a well-defined visual twist; there was no evidence whatever of any fatigue or prior cracking. Microspecimens cut from close to the fracture and remote from the fracture showed steel of a satisfactory standard regarding cleanliness and a metallurgical condition without any problems. Hardness tests showed that the material was within the tensile range which could be expected for a component of this type.

These results were discussed with the client and it was agreed that with all the evidence present there was no need to carry out any further examination such as mechanical testing, chemical analysis, etc.

This is an example of failure investigation which had the mechanism clearly defined and in discussion with the client the problem was resolved without any expensive or confusing additional work.

GROUP 4 CORROSION/CONTAMINATION FAILURES

These case histories involve problems of corrosion and contamination. Corrosion is a chemical attack on the surface of metals. The most common corrosion is oxidation of steel identified as rust. Pitting, discoloration, reduction in section, and presence of corrosion products also identifies corrosion.

While stress corrosion is now commonly associated with failure, it is not often identified as a 'mechanism', as it generally initiates fatigue, which is then found at the subsequent examination.

Corrosion failures can have the corrosion products present, or they may have been removed by service conditions – or cleaned away by the client. Subsequent work often requires chemical analysis, and the problem of cross-contamination must always be considered.

Contamination will often appear at first sight to be the result of corrosion, but cleaning and visual examination will clearly differentiate between them. Contamination may, however, present long-term problems and the investigator must be aware of these potential hazards.

Corrosion failure of an austenitic stainless steel toilet bowl

This was the result of weld decay. Discussion identified that carbon electrodes were used to tack weld prior to inert gas welding.

Leakage occurred on a high quality austenitic stainless steel toilet bowl which had been in service for only a short time. Failure was confined to three points and investigation showed that this was the result of weld decay.

Weld decay is caused by chromium carbides migrating to the grain boundaries when austenitic stainless steels are held at temperatures between 300 and 600°C. This gives a two-phase structure which corrodes. If the carbon content of the steel is held below 0.05% then there can be no chromium carbides. The steel specified and used had such a low carbon content.

The reason for this problem was identified in discussion with the makers, where it was admitted that for expediency, and without knowing the significance of the action, the components were assembled in a jig and then tack welded. A number of components had been tack welded using carbon electrodes. Once this was identified, all others in the batch were withdrawn before any other failures took place.

Corrosion failure of a cadmium plated shaft in a bronze bearing immersed in sea-water

This is an example of the wrong design specification.

This component was part of a bogey used in a marine operation involving a rail track which ran below the water to allow vessels to be lifted out of the sea. It had two axles. Once the vessel was fixed to the bogey it could be pulled up the rails and remedial action carried out on the vessels above high water.

The bogey was manufactured, tried and shown to be operative, but then left static for 3–4 months. When it was next required it was found that all axle bearings had seized solid.

Examination of these, along with the specification, showed that the shaft had been designed in mild steel, to be cadmium plated, with the bearing in bronze. The cadmium plating was estimated to have a thickness of 0.01 mm (0.0005″). This plating had failed completely where it was in contact with the bronze, as the sea-water had acted as a 'lubricant' and as an electrolyte. When the cadmium disappeared through galvanic action or chemical attack, the substrate steel corroded, with the corrosion products seizing to the bronze bush.

The report was highly critical of the use of cadmium plating or any other electrodeposit under these aggressive conditions. It was stated that the shaft should either be manufactured from some form of stainless steel or have had a very positive protection such as PTFE coating.

It was pointed out that while cadmium has a reputation of being better than zinc for marine purposes, the improvement of one over the other is marginal, and no metallic coating of the thickness specified, apart from probably gold and possibly nickel, would have withstood the arduous conditions of being immersed in sea-water and then allowed to air dry whilst in contact with a bronze.

Apart from any chemical attack, which was certainly the main cause of the problem, there was the possibility of galvanic effects of cadmium and bronze. There is a voltage of approximately 0.45V between cadmium and copper.

This is an example of the poor quality of information used or supplied to a design engineer.

Corrosion and non-corrosion on identical components under identical conditions

These two case histories were investigations several years apart. The first was of mild steel containers used in food processing, the second modified mild steel anchor chain used on offshore oil rigs.

In both instances one batch of apparently identical steel had serious corrosion without any change in the operating conditions, resulting in its being scrapped.

The first investigation involved mild steel baskets, approximately 1 m diameter, 1.5 m high with 50 mm holes. The baskets were filled with food containers and loaded into autoclaves filled with chlorinated water brought almost to boiling.

After almost twenty years an increase in demand required 20 additional baskets to be supplied. With less than six months usage, severe pitting corrosion was found on the new baskets.

The second investigation was of 70 mm diameter anchor chain. Two lengths, each over 1 500 m long, manufactured within six months of each other, were stored side by side as spare lengths of new chain in the upper reaches of the Tay estuary.

Both lengths were removed at the same time and one was found to be satisfactory, while the second had serious pitting corrosion and was adjudged by all concerned to be unacceptable and thus scrap.

Investigation of the baskets found that the originals had been made from hot rolled steel sheet, while the new baskets were grit blasted to improve the appearance. In this instance the new baskets were salvaged by heating to 650–700°C for several hours. These baskets are still in service.

The anchor chain problem required specialist knowledge on procedures used at manufacture. The investigator commented that the Tay is a fast flowing, aerated river estuary that changes from fresh to saline every tide. It could thus be expected to corrode steel, and should not have been used for chain storage. The chain which corroded was manufactured, heat treated, then grit blasted, inspected and tested before being stored.

The satisfactory chain was manufactured, grit blasted, inspected, then heat treated and tested. It is thus suggested that the oxide film produced by heat treatment acts as a corrosion protective film where some corrosive conditions exist.

This illustrates the difficulty the corrosion investigator can encounter, as the difference between corrosive and non-corrosive metal surfaces and atmospheric contamination can be significant when only slightly different processes are involved.

Blockage of diesel fuel filters resulting from bacteria contamination

Blockage was identified in the filters of fuel tanks on diesel ferries. It was found that this was bulky iron hydroxide/iron oxide, and examination of the fuel tanks showed that locally there was corrosion near the base of the tank. The tanks were examined while empty but visual examination indicated that the corrosion was present only at the level where moisture could be trapped, with the diesel floating on the surface.

It could be seen that there was evidence of quantities of iron oxide of various colours from bluish green to black to reddish brown. When these were removed, some pitting was found to an insignificant degree.

Samples were collected and examined in the laboratory. The investigator had indicated that the most probable source of the problem was contamination from bacteria which can exist in brackish water and it would appear that iron can act as a catalyst to allow bacteria to produce copious amounts of organic iron compounds.

The laboratory examination confirmed the presence of organic matter in considerable amounts associated with the iron oxide, and no further investigation was carried out in this area.

Attention switched to the fuel tanks. The problem was that the water, which must exist in a tank of this type because of condensation, was not being allowed to drain to the lowest point. There was an absence of ‘mouse-holes’ at the corners where the metal struts (to prevent surge of fuel) met the side walls. This meant that as the moisture condensed, there was a much larger area of moisture/fuel interface. This was pointed out to the operators, who identified that this was a mistake in construction, and once this was remedied no further problems occurred.

The remedy worked well because the interface area was much smaller and the procedure for draining off moisture resulted in no further fuel contamination.

This is an example of the result of visual examination being confirmed by chemical analysis, and discussion.

Following this investigation and report, the client carried out a survey of other vessels. It was found that the organic/bacteria contamination was not uncommon at the fuel/diesel interface. On the other vessels the feed pipe to the filters was always well above this interface, and thus did not contaminate the filters.

Corrosion failure of pipe connectors on tankers, caused by use of wrong material

The upper deck of tankers has a myriad of pipes, all of which must be connected. The connections normally use type 316 stainless steel, but naval brass has also been found to be successful. This is a standard brass with approximately 1% aluminium which significantly improves the corrosion resistance under saline conditions. The client involved had considerable successful experience of this material.

On one ship, however, it was found that fracture and leakage occurred on certain connections when they were adjusted.

Examination showed that fracture was by brittle tensile failure from intergranular surface corrosion. Further examination and chemical analysis showed that the material did not contain aluminium and therefore was not naval brass.

This is an example of a simple investigation where the answer was supplied rapidly and efficiently with the minimum of work and resulted in no serious consequences except that in future a certificate was to be obtained with all naval brass components.

The experienced metallurgist can see the difference in colour between naval and standard brass. The naval brass with 1% aluminium has a more definite yellow colour than standard brass. The aluminium content stabilizes the surface oxide, giving much better corrosion resistance.

Corrosion of nitrided 12% chromium stainless steel

Two investigations have involved the failure of 12% chrome steel which had been nitrided and was then used in aggressive corrosion conditions.

Firstly, a die plate was involved in the final operation of a plastic extrusion process. Problems had been identified with corrosion and the specification was changed from high tensile steel in the hardened condition to 12% chromium steel. This gave a better life but was found to lose the sharp edge after a very short time. It was therefore suggested that the component should be nitrided to give the corrosion resistance of 12% chromium steel with the high hardness of nitrided steel.

This was carried out on a limited number of die plates and it was found that the results were disastrous. After a very short period of time the cutting edge was removed, but worse than that the mating faces were badly scored. Investigation showed that the cutting edge was failing rapidly, resulting in debris with high hardness (above 900 DPN) being trapped between the two mating faces causing scoring. Examination of the mechanism showed that this was corrosion of the contact surfaces, resulting in metal plucking out.

Further tests showed that when 12% chromium steel is nitrided, it loses its stainless properties.

The problem was reduced by increasing the hardness of the 12% chrome steel. Further work was planned to produce dies with the cutting surfaces induction hardened. The process, using the dies, was superseded before any trials were carried out.

The second example investigated was of underwater control mechanisms where for corrosion reasons the plungers were specified in 12% chromium steel. It was found, however, that they were wearing. The answer appeared to be to nitride the plungers, and this resulted in a similar problem to that discussed above, that is, the production of very hard abrasive particles which then increased the clearance. In this case it was more difficult to identify the reason for failure as by the time the unit was investigated there was considerable wear, and evidence of the corrosion had disappeared.

It is interesting that in both instances the corrosion was confined solely to the nitrided layer. Once this had been removed, either by corrosion or abrasion, then the remainder of the surfaces were satisfactory apart from scoring. The resultant clearance between the working faces could not be accepted.

These investigations identified the problems, but did not result in the investigator being able to suggest a reasonable cure. There remains no economical cure to abrasion/wear, where aggressive corrosion is present. This invariably requires expensive, corrosion-resistant, tough hard alloys such as cobalt or titanium alloys.

Case histories where stray direct electric current has resulted in corrosion

Where a metal is made anodic in a conducting cell with an electrolyte, then corrosion can be expected. This corrosion will occur whether or not the material is corrosion-resistant under normal circumstances. For example, gold or titanium can be corroded if they have a sufficient direct electric current applied to make them anodic to another conductor in a conducting fluid. Examples of this follow.

(a) Aluminium fins on copper tubes used for heat exchange purposes are commonly used where the liquid is passed through the copper tube, and air is blown across. With the copper tubes alone the heat exchange is quite low, thus inefficient. With fins present there is a very much larger surface area for the heat to be removed, firstly by conduction to the fins and secondly by convection as the air is pulled across. Aluminium fins have been found to be more economical, and just as efficient as copper fins.

In theory galvanic corrosion should occur between the aluminium and the copper, with the aluminium corroding. In practice it is found that only under certain conditions does corrosion occur, and this corrosion is normally from the tips of the fins, not at the roots where galvanic corrosion would occur.

It was found, however, that a large heat exchanger in a slightly corrosive atmosphere which had been protected by chromating the aluminium had two surfaces of the heat exchanger corroding seriously, with the opposite two surfaces in perfect condition.

Investigation using a millivolt meter showed that where the corrosion occurred there were 200–400 millivolts across pipe and fin, making the copper pipe with aluminium fins anodic to earth. With the aluminium being the more anodic of the two, serious corrosion from the aluminium occurred. This was cured by ensuring that all the aluminium finned copper tubes were fully earthed. The investigator suggested that the source of the stray current should be identified, but this was not achieved.

(b) The second example was the end plate on an air receiver. This was insulated from the remainder of the receiver firstly by a non-conductive gasket, and secondly by the use of PTFE washers below the head of the bolts. When this was inspected it was found to have bright pits on the inner surface and it was suggested that the only explanation of these bright pits was electrolytic corrosion.

Tests were carried out using a recording millivolt meter with one probe on the end plate, the other probe on the body. It was seen that the body had no voltage applied, whereas the end plate had a voltage of up to 120 millivolts. There was no obvious reason for this but the problem was cured by using standard steel washers in place of the PTFE insulating washers.

(c) The third failure was of studs inserted into concrete where a wooden template had been used to locate the steel studs in the concrete. This template was left in position for a period of months and on removal it could be seen that the steel studs showed evidence of very severe local corrosion. This could only really be explained either by the wood being extremely acidic, which simple tests showed was not the case, or electrolysis.

Tests with the millivolt meter showed that the studs were 200 millivolts positive to a good earth. It was then discovered that these studs were in contact with the reinforcing bar of the cement, and this reinforcing bar was also 200 millivolts positive to a good earth.

Further investigation identified that the reinforcing bars was unintentionally connected to the impressed current corrosion protection circuit applied to an area on the construction site. This uses inert anodes positive to the unit being protected. In this case the impressed current inert anode had become connected to the reinforcing bars in the concrete, resulting in the corrosion found.

This effect was cured by earthing all components as it was considered that this was simpler than locating where the inert anode was in contact with the reinforcing bars.

The investigator did not agree with this technique as it was felt that other problems could arise. To date the system has been successful. It would, however, have been much more satisfactory to identify where the connection between the inert anode and the reinforcing bar existed, and to insulate this connection.

Problems of heating austenitic stainless steel, or when contaminating the surface

Austenitic stainless steel has a nominal analysis of 18% chromium, 8% nickel. It thus has more than the magic figure of 12% where a chromium oxide forms in lieu of iron oxide. This chromium oxide is coherent and adhesive and forms immediately it is abraded if in an oxidizing atmosphere.

There is a history, however, of iron contamination from austenitic stainless steel vessels. It has now been shown that this is the result of either heating in an oxidizing atmosphere or more commonly, welding, where obviously heat must be applied.

When heat is applied above approximately 500°C to austenitic stainless steel in an oxidizing atmosphere, the chromium will preferentially oxidize and thus denude the surface of chromium. Once the chromium is below approximately 9% then it no longer supports chromium oxide on the surface.

The problem was identified on a massive scale originally with the Scotch whisky industry. When production increased and it was decided to bottle whisky in the country to which it was exported, rather than the country of origin, large stainless steel tanks were manufactured and used to transport whisky in bulk – up to 5,000 gallons at a time. The first container to arrive was found to have green whisky, which is not normally acceptable!

Investigation showed that this was the result of oxide forming associated with the welds. This was not visibly identifiable but the whisky attacked this area, and approximately 20 parts per million iron in whisky will turn it green.

Other areas where this problem has existed are, for example, where blasting is carried out with austenitic stainless steel or even where shot blasting is carried out adjacent to stainless steel. This can result in a fine film of iron oxide either from the debris of blasting or from the blast material, which very often is cast iron or steel shot. This will eventually rust and is normally unacceptable.

There is a test for this, known as the ferroxil test. This uses potassium ferricyanide which is a light green colour but which immediately turns to a dark Prussian blue when in contact with iron. This is a very sensitive test and requires some skill to get sensible results.

Reduction in thickness of hot water copper pipes – offshore installation
It was identified that leakages were occurring frequently in the hot water system on a North Sea oil production unit. Samples sent for examination showed no evidence of any obvious material defect. The leakage appeared to occur randomly by thinning of the copper tube which then burst.

A visit was made to the production unit and there the problem was examined by crawling along the roof space where all the service pipes, etc. were fitted. It could be seen that the leakage was confined to certain areas and temperature checks identified that these areas had a higher water temperature than other locations.

This was discussed and it was found that the thermostat controlling the water temperature had been adjusted to be approximately 10°C higher than specification in these areas. After considerable debate it was found that shortly after the unit was commissioned, the water temperature for ablutions dropped too rapidly, thus the temperature was raised.

Further discussion took place onshore with the designers and maintenance personnel. This elicited the information that the capacity and temperature of the hot water system was designed for a specific number of operators, and using a standard formula the capacity had been calculated, and resulting energy input installed. What had not been appreciated was that the usage of hot water was not on a regular basis. Four times in 24 hours, 0615–0645, 1215–1245, 1815–1845 and 0015–0045 hours, there was a very high water demand. This was when the shift system in 25% of the workforce stripped and had showers. Those at the tail end of the queue at each shift complained of lack of hot water and this had resulted in the increase in thermostat setting whereby those who had their ablutions first had very hot water and therefore used less hot water by diluting it with cold water. The increased temperature resulted in an increase in attack of the copper pipe, thus reduction of the wall thickness and some increase in pressure.

Following the report, a larger header tank for heating water at a lower temperature was installed and leakages no longer occurred.

This is an example of a limited amount of investigation at a low technical standard followed by discussion and the application of common sense.

It also shows the critical nature of corrosion where a relatively slight increase in temperature resulted in corrosion in months which would have not occurred in years at the design temperature.

GROUP 5 ABRASION/EROSION FAILURES

This group covers abrasion and erosion. There can sometimes be initial difficulty in differentiating between corrosion and abrasion or erosion. Abrasion/erosion is mechanical, corrosion is chemical.

Once the surfaces are clean, the investigator should always be able to differentiate the pitting, etc. of corrosion from abrasion or erosion by visual examination.

Cavitation erosion may present the greatest difficulty, but with experience and knowledge of operating conditions the possibility of cavitation will be appreciated and the characteristics of the pits recognized.

Abrasion failure of economizer tubes

This resulted from grit in the flue gas of a boiler installation.

These tubes were for boiler feed water heated by the exhaust combustion gases. Unscheduled failures were identified on a number of tubes, always at the hairpin bend in a similar location. It was assumed that this was the result of corrosion, and discussions were taking place to have all the tubes replaced with more expensive material.

An investigation was requested which showed that the failure was not corrosion but abrasion. This was identified at visual examination.

Micro-examination showed no evidence of corrosion. The surface was being abraded, with evidence of cutting but no cold work on the surface. Further investigation found that there had been a failure of the dust arrester locally, which had allowed some of the abrasive particles normally removed from the gas stream to escape. The velocity of the hot gas stream had contributed to abrasion.

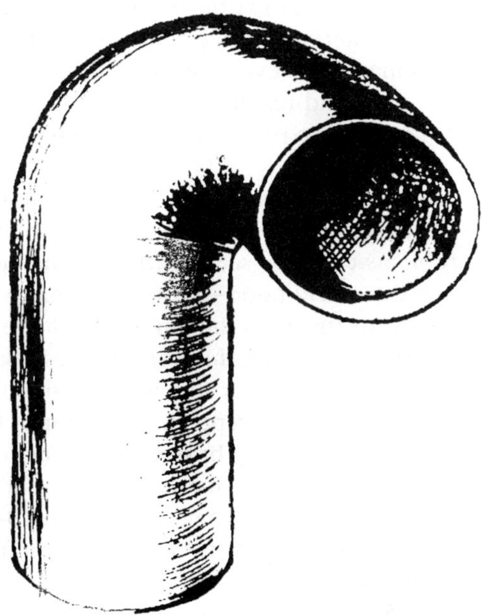

Figure 15.11 Section of the economizer tube, showing the dramatic thinning of the tube wall on the outer surface exposed to abrasive action.

This is an example of the value of failure investigation. The opinion of those involved with the equipment was that the problem was corrosion and steps had been taken to change the material at considerable expense, whereas the correct information showed the problem to be abrasion from an external source.

In this instance, the remedy was simple and was adopted. The more expensive corrosion resistant tubes would probably have failed even earlier than the plain carbon tubes.

Failure of chain caused by abrasion

This resulted from a very poor surface finish on loaded pins.

This investigation was the failure of a bicycle-type chain link with a pitch of 150 mm and a pin diameter of 50 mm. Failure had occurred with an approximate 75 ton load applied, resulting in considerable damage.

Investigation showed that the mechanism of failure was abrasion, with the carburized pin wearing its way through the carburized bush. Once the bush had worn, the pin continued to wear the soft link until tensile failure occurred.

It had been reported that the chain was extremely noisy in operation and the manufacturers had repeatedly recommended better greasing. The investigator found no shortage of lubrication, but the grease contained considerable quantities of abrasive particles produced by the abrasion of the hardened bush against the hardened pin.

Examination of new links with the bushes in the 'as received' condition showed that these had an internal surface finish somewhere in the region of 120 micro-inches. This meant that even with satisfactory lubrication, loading, particularly with impact, would result in fracture of the brittle peaks. Once these fractured then the abrasive particles became trapped in the grease and acted as a lapping mechanism.

Micro-examination and hardness tests showed that the bush and pin had a satisfactory carburized case with an adequate merge between case and core, and no problems could be seen apart from the fact that the surface finish of the bush was extremely poor – typical of what would be produced by drilling. The standard of finish on the pin was in the region of 20 micro-inches and while some wear took place, this was much less than that of the bush.

It was suggested that the reason for this was the smoother finish on the pin. The abrasive particles were not trapped. On the bush, however, because of the poor finish, the particles as they broke off became trapped and caused abrasion. It was recommended that for immediate remedial action, the bushes should be ground to give a surface finish no worse than 10–15 micro-inches. Whilst this would result in some play between the bush and the pin, provided the chain was correctly loaded this should not cause any problems.

This illustrates the importance of visual examination with the necessary experience to realize that a 120 micro-inch finish is not acceptable for hardened bearings.

Unfortunately the conclusion of this report was not accepted, and the use of bicycle-type chain was discontinued. The investigator could not persuade the client to accept the technical conclusion, and an expensive alternative remedy was installed.

This is therefore an example of the frustration of the independent investigator who has no legal or commercial clout, but is certain that his client has made a wrong decision.

If the consequences are only financial the answer is silence. If there are health or safety implications, the investigator must arrange to publicize the findings. If there are any political implications, the investigator should not become involved!

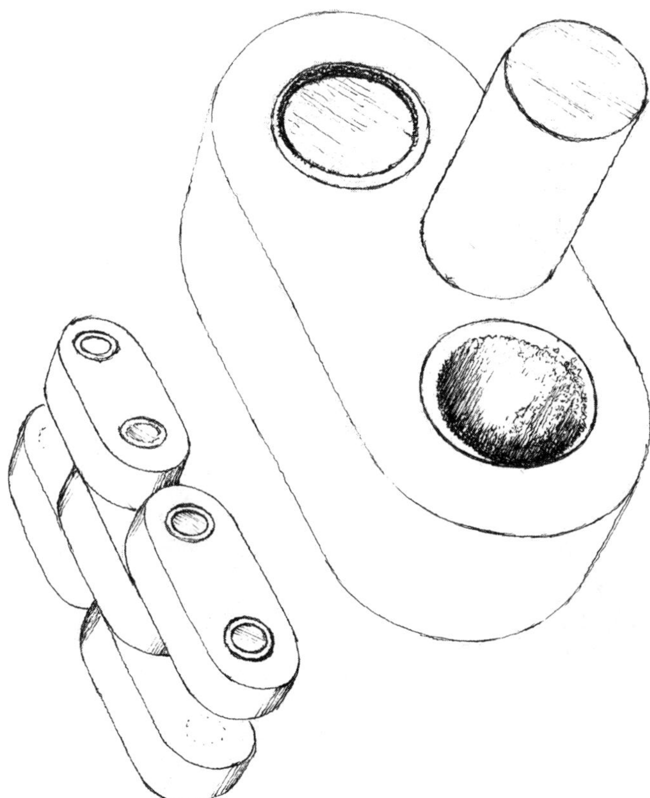

Figure 15.12 Bicycle-type chain failure. Enlarged link shows the rough surface finish of the bush and the worn area of the bush on the loaded side. Once the bush is worn, the softer link rapidly fails.

Cavitation erosion on water-cooled heavy-duty braking system

The component involved is a rotating mild steel brake on a large 2 metre winch. The braking resulted in high temperatures, and a cooling system had been devised with an annular space on the outer surface through which cooling water was passed.

Because of corrosion problems, the annular component was maufactured in relatively thin austenitic stainless steel which was welded to the surface of the brake.

After a relatively short time in service leaking was found of a serious nature. The leaks were not associated with the welds but with certain portions of the steel annular component.

When this was examined, it showed that the inner surface of the mild steel component at specific areas around the annulus had been subjected to cavitation erosion. Some of the cavitation pits had penetrated the surface, allowing the leakage to occur. The surface of the stainless steel was not affected.

Micro-examination showed no problems with either the mild steel or the austenitic stainless steel which could have contributed seriously to the cavitation. In discussion and on trials, it was seen that when braking occurred there was considerable vibration and in some instances resonance.

It could be seen that the vibration was consistent all round the annulus, whereas the resonance, which could be seen by vibration of the steel annulus, took place at specific areas which were related to the stiffening of the mild steel brake drum.

Cavitation erosion is the result of pressure of a fluid on a surface which is suddenly reversed towards a vacuum. That is, there is a cycle where the

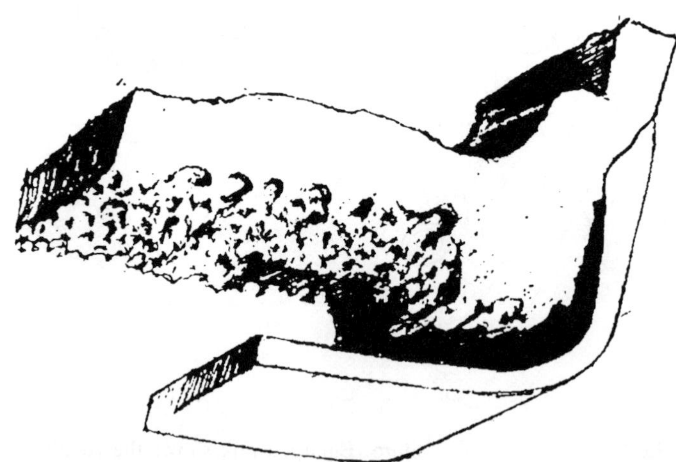

Figure 15.13 Rounded pits typical of cavitation erosion.

fluid pushes the surface of the metal, and on the reverse cycle, a suction is applied to the surface of the metal. Some metals are capable of withstanding these forces, others are not.

It is now known that the cavitation process plucks out the surface grains, and that it is metals or alloys which have strong grain boundaries which best resist the problem. In this instance the mild steel had large grains with inherently weak grain boundaries. The stainless steel subjected to exactly the same loading had ductile grain boundaries.

This is an example of a failure which, while not unique, is rare and required the experience of the investigator to identify cavitation erosion.

Cavitation erosion is a very specific failure where rounded pits are formed which are smooth and immediately after the action of cavitation are bright.

The pits are vertical to the surface and are thus quite different from normal erosion, where the pitting or channels produced are at an angle to the surface. They are also different from any other form of corrosion in that they are generally very homogeneous and invariably smooth and rounded, whereas normal corrosion has a rough surface.

Cavitation erosion of cylinder liners from resonant vibration

This failure occurred on high performance internal combustion engines with wet liners. The liner had a thin wall section sealed at the top and bottom, and the coolant circulated around the cylinder to remove the heat from combustion.

With some engines used at peak performance, holes appeared on the outer surface of the liner at local points on the circumference. These were tiny holes, but eventually they perforated the wall, pressurized the coolant system and also sucked coolant into the combustion chamber. This resulted in 'hydraulicing', with severe damage to the connecting rods and cylinder head.

Visual examination showed that the holes were clean, with no evidence of corrosion. The walls were parallel and the holes originated from the outer surface of the liner. Further examination showed that the bottom of the pits which had not perforated the section were always bright. Micro-examination through these showed that the metal grains were being plucked out, that is, it was a tensile type failure.

It was therefore identified as cavitation erosion in which the surface was being subjected to a push/pull action.

Further tests showed that at certain revolutions with full power, resonance took place at local areas around the circumference, giving a very high frequency and resulting in the push/pull action necessary for cavitation erosion.

The problem was eliminated firstly by controlling the revolutions, and then by altering the profile of the wet liner to alter the resonance points.

The mechanism of cavitation erosion is described in the abrasion section of Chapter 7 on Mechanisms of failure.

GROUP 6 HEAT FAILURES

This is where the mechanism is seen to be associated with heat. This can be damage, for example to paint, fabric, plastics/polymers, which can occur at relatively low temperatures – below 300°C.

Failure of metals can also occur at these temperatures over a period of time and will not be found visually as a mechanism of failure but should be identified at the micro-examination.

Visible heat damage to metal at higher temperatures can be discoloration, distortion, scaling or melting, but with steel and most other metals in common use, temperatures above 500–600°C are required to show any visible change.

Fire damage to steel structures

This can cause distortion, and in rare instances metallurgical damage. A number of such investigations have been carried out, resulting in non-metallurgical inspectors being able to assess the possible damage.

These investigations were firstly to identify the extent of the resulting metallurgical damage, and if possible to comment on the reason for the fire initiation.

In none of the investigations involved was there any evidence that the heat of the fire had resulted in metallurgical changes to the metal. There was, however, evidence of heat distortion caused by the local expansion of metal joists and columns.

The heart of the fire could be identified by the extent of the metal expansion, and visible evidence of the extent of damage or combustion to other materials, in particular organics such as plastics, wood and fabrics, and to a lesser extent plaster, brick and concrete.

The metallurgical investigator is aware that paint, plastics and organic materials are affected by discoloration, charring, blisters and distortion at much lower heat levels than that required to have any effect on metals.

The effect of smoke can supply meaningful information in some instances. Smoke can, however, affect areas which are remote from the source.

Visual examination will identify where the maximum heat occurred. Thin metal panels or trim can be expected to distort, particularly aluminium, which has a higher coefficient of expansion than steel. If metalwork has any trace of paint or organic matter remaining then it can be reported that no metallurgical change has taken place.

The necessary temperature to soften or harden steel is above 600°C, which will destroy any organic material. If the metal is naked and shows visual evidence of heat a micro-specimen or specimens will be cut by hacksaw. In some circumstances the investigator would also prepare a sample from the same type of metal remote from any heat affected material.

The specimens are examined unetched, then etched, for grain growth, excessive surface oxidation, and in extreme cases local hardening.

The author has been involved with numerous fire damaged buildings, and has now educated the clients, mainly insurance companies and assessors, that if any paint, oil or organic type matter remains then there can be no significant metallurgical changes.

Engineers can identify and assess the significance of any heat distortion where this has occurred.

Investigation of heat damage resulting from a fire raising incident

This was an investigation carried out following a fire in a company producing wooden components. The fire fighters themselves, and the fire assessors, were suspicious regarding the amount of damage and the assessors were very perturbed at the size of the claim which had been submitted.

The building had been sealed and was investigated by a number of people, including the author. Visual examination revealed immediately that while some damage existed, particularly of the structure, where smoke had caused considerable discoloration and there were some heat effects on organic type materials, this was to a limited extent.

The equipment where claims had been made for damage by distortion could be seen to be expensive and well-maintained, but the visual examination showed that there was oily sawdust remaining on some of the areas, and others had dry wood cuttings.

The seat of the fire could be seen to have been one unit and it appeared that this produced smoke which acted to open the cowling on the roofs, allowing release of the smoke, and a neighbouring factory immediately called the fire brigade. Their rapid response resulted in extinction of the fire and little structural damage.

At the insistence of the client, that is, the insurance company, samples were taken of paint and oily sawdust and experiments carried out regarding the amount of heat which would be required to cause the problem, either on the paintwork of the steel or the oily sawdust.

These showed that temperatures as low as 100°C would cause some of the problems, and the maximum temperature to cause the paint distortion was less than 300°C.

Figures were then produced showing the coefficient of expansion of metals at these temperatures, and this refuted information claiming that the heat had caused distortion resulting in damage and necessitating expensive remedial action to the equipment.

Fire damage to hanger roof, resulting in visual damage without distortion

An aircraft hanger built in the late 1930s, now used as a store, was subjected to some local heat damage by a fire adjacent to the building.

The investigation was to identify any distortion and to quantify the degree of damage for insurance purposes.

Examination showed that the roof was made from sheet metal approximately 10 mm thick. This had been prepared for long-term corrosion protection by blasting and the application of red lead primer to a high standard, followed by two coats of bitumen paint. This was identified by samples taken adjacent to the damaged area. The damaged area, where the paint had been destroyed and the roof was rusting, was less than 1 metre wide.

There was then an area where the bitumen coating had softened and had been destroyed locally, leaving the red lead intact. The remainder of the roof was found to be in pristine condition after more than 60 years of service.

It was recommended that all that would be required would be to locally blast the naked surface, the surface with exposed red lead and the blistered surface. It was pointed out that some scraping might be necessary with the bitumen and that all operators involved in the process must wear positive protection against any dust from the red lead.

It was then specified that some form of epoxy paint with a positive primer should be applied. It was also suggested that it was unlikely that this portion of the protection would last for another 60 years. However, a product correctly prepared, correctly primed and with sufficient thickness of paint will have an adequate life.

It is now generally agreed that there is no excuse for paint failure except under the most aggressive conditions or where damage occurs, provided the surface is correctly prepared, primed and good quality paint applied under controlled conditions. There are equivalents to red lead, as this, quite rightly, is now a banned substance.

Failure of architectural aluminium trim on a superstore

This investigation was to find why aluminium extruded sections used for architectural decoration collapsed. It was discovered that the trim on one side of the building had fallen off.

Investigation showed that failure appeared to be the result of expansion of the aluminium, this expansion resulting from the sun shining on this side of the building.

Further investigation and discussion found that the design specification required a space to be left at the end of each section of the trim. This had not been done, so each section butted against the next, having been left a space only at the end. While this could not be seen on the section which had failed, the other three sides of the building where the sun had not shone on the surfaces all had the sections of trim in contact with each other with a gap between the end of the length of the section and the edge of the building.

Calculations showed that aluminium, which has a high coefficient of linear expansion, shows a significant increase in length with a temperature increase of 10°C. As expansion could not take place because of the friction between the sections, when the section tried to expand it fell off.

This is an example of a design specification being correctly produced but not followed in production.

GROUP 7 HISTORIC FAILURES

This section contains information on a number of dramatic failures which have some historic, as well as technical, interest.

The failure which probably advanced the 'science' of investigation most is included.

The author wishes to record here his own theory that it was investigation of aircraft failures that radically changed the historical 'cure-all' approach of making things bigger rather than finding the reason for failure.

Paddle steamer 'Waverley'

The Waverley at approximately 50 years of age is the world's oldest ocean-going paddle steamer, and still operates around the coasts of Britain. The original boilers were the 'Scotch' type, which have the ability of operating under adverse conditions. These were replaced approximately 15 years ago with the modern tube type boiler. The boiler is required to generate steam to drive the reciprocating pistons for the two paddle wheels.

Failure of the new boiler occurred with loss of steam. The 20 mm thick boiler plate distorted under pressure from the water side, reducing the metal thickness to less than 1 mm.

Investigation identified firstly that hydrocarbons, that is, oil-like substances, were present on the water side of the boiler. Micro-examination showed that the boiler plate was of a standard modified mild steel and that at the thinnest portion, before rupture occurred, it was estimated the temperature had been in excess of 700–800°C and might have been as high as 900°C.

At these temperatures steel has a low tensile strength and thus the pressure of the water/steam would stretch the steel, resulting in the failure identified.

It was reported that the reason for the high temperature was an oil film on the water surface of the boiler. There was thus a sandwich of steel, oil, steam and water. The combination of the layer of steam and oil would dramatically reduce the heat transfer to the water and thus increase the temperature of the steel. As the steel temperature increased, the steam would increase until eventually the boiler at this section would become red hot. The temperature would then increase as there would be little or no transfer of heat from the water side of the boiler. Rupture would then occur.

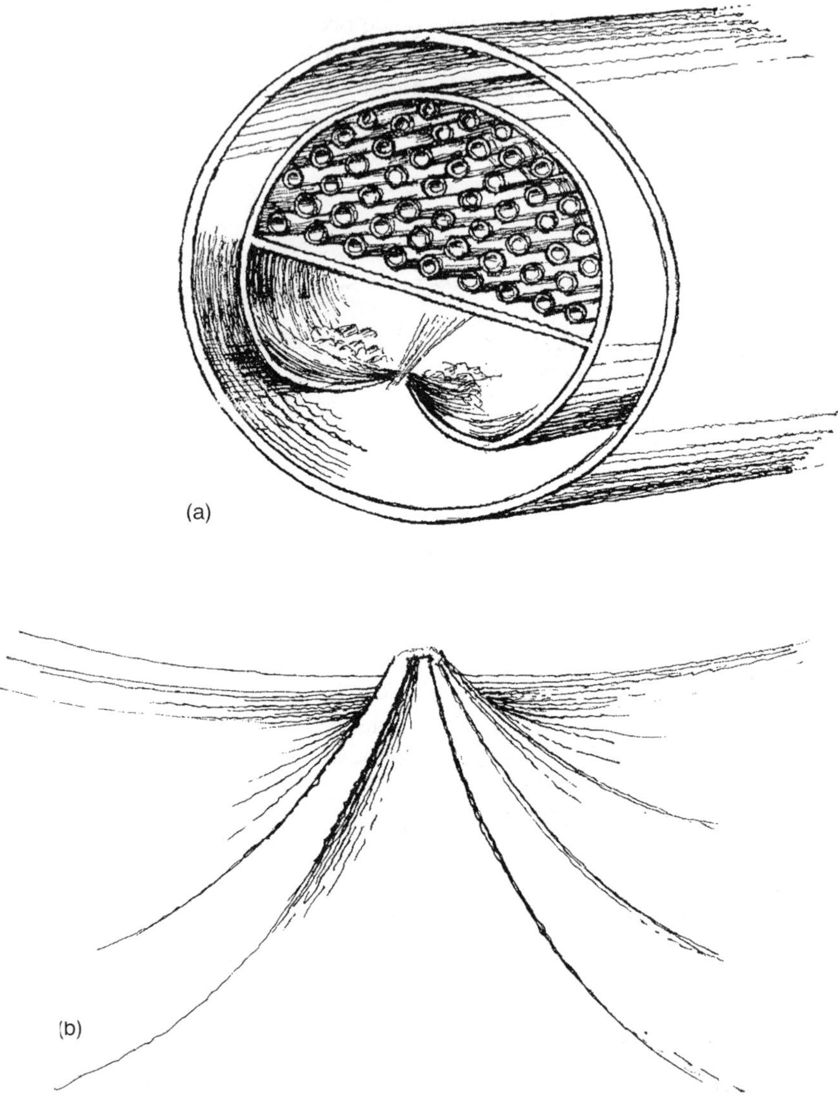

Figure 15.14 (a) The boiler. The water/steam jacket is between the outer and inner vessels. The flame chamber is at the centre, below the water tubes. (b) An enlargement of the point of failure.

The accepted conclusion was that the failure was oil contamination inside the boiler. This was thoroughly cleaned and the Waverley went back to sea with thermocouples fitted inside the boiler recording the temperature of the steel.

After less than 100 hours sailing, it was identified that high temperatures existed on local spots within the boiler on the water side, and she was withdrawn from service.

The situation was discussed when it was identified that the traditional tannin-based water treatment system had been replaced with a more sophisticated procedure. The engineers were delighted with the new system as no contamination problems with the oil appeared to exist. It was, however, pointed out that the oil would still be entering the boiler and instead of floating to the water

Figure 15.15 Aerial view of the paddle steamer 'Waverley' (as would be seen by the captain as he proceeded upwards if the boiler had exploded).

surface, the new modern treatment, which is very satisfactory for boilers which do not have oil contamination, would result in a film of oil forming on the recently cleaned boiler surface.

While no experiments have been carried out to prove or disprove this theory, the Waverley now happily steams around Britain with the old-fashioned tannin boiler treatment and the oil manually skimmed off periodically.

This is an example of a failure investigating identifying rapidly the mechanism of failure without initially identifying the reason for the failure. This was found by discussion, not by investigation.

The Comet aircraft

The Comet aircraft led the way in jet aircraft passenger service. This was a UK plane of considerable artistic merit, quite efficient, and is still operating.

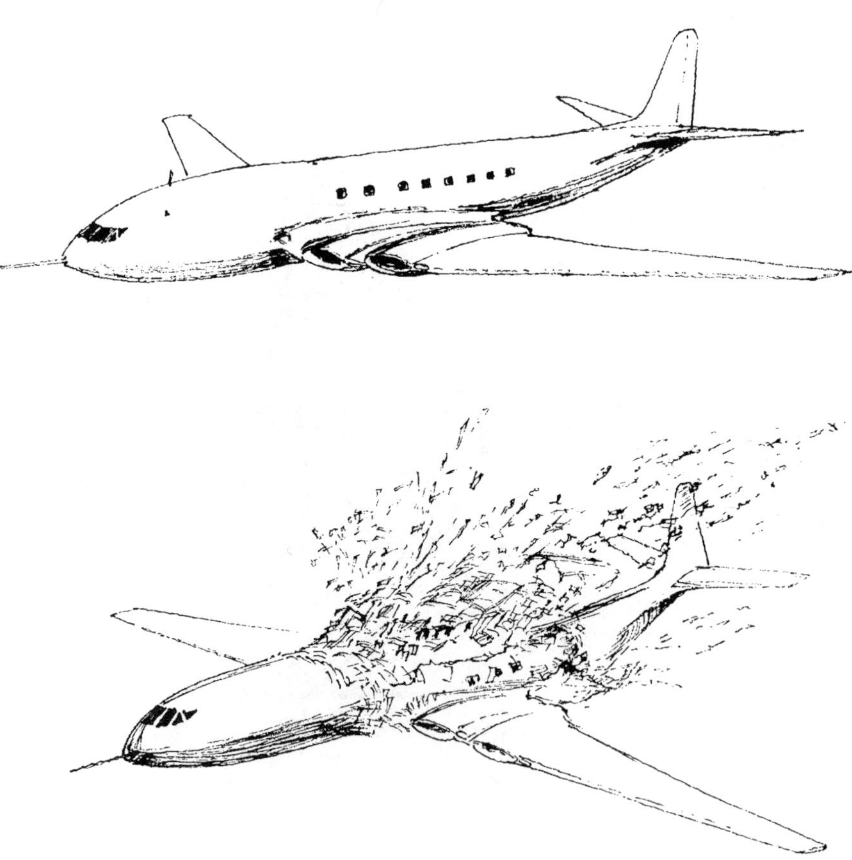

Figure 15.16 The Comet air disaster.

Three aircraft crashed, however, with considerable loss of life. The first came down over mountains and it was presumed that turbulence caused the failure. There must now be doubt regarding the reason.

Both of the other aircraft failed after leaving Rome, and when the third failed the Comet was grounded. The fuselage was recovered at great expense, and this led to a classic failure investigation where the Civil Aviation Authority supervised the rebuilding of the fuselage. There is little doubt that this investigation furthered the world's knowledge of fatigue failures.

It was identified that a relatively simple defect resulted in a stress raiser initiating fatigue such that every time the aircraft went above a certain height, the pressure inside the fuselage applied a tensile load for fatigue propagation. This eventually resulted in the fuselage failing, and with the increased pressure inside, the contents of the cabin were sucked out and the aircraft demolished.

The resulting report revolutionized the knowledge of fatigue initiation and propagation, with test procedures being required on all new design aircraft, and considerable information was made available to reduce fatigue stress raisers.

Sinking of the warship 'WASA' in 1640

A case history of poor communication.

This ship was to be the pride of the Swedish Navy. It had various modifications at a late stage in building to make certain that it was a better ship than any the Danish or English navies had. These included adding additional guns, so increasing the top weight.

It left the shipyard, sailed across the harbour of Stockholm and in an 8-knot wind fell over and sank, with flags flying and sails set!

An enquiry established that several months previously trials had been carried out with 30 men running backwards and forwards across the deck. Before they had managed four crossings the ship was keeling so badly that the Admiral who had made the request stopped the tests but did nothing else.

The ship has now been recovered and is a magnificent exhibit in a museum in Stockholm where it can be clearly seen that the ballast, which is less than head height, was all that kept the ship upright.

While this has little or no metallurgical significance, it is a very interesting failure which would not have happened if cognisance had been taken of the results of a simple test.

Tay Bridge disaster caused by poor workmanship

This resulted in the magnificent but over-designed Forth Railway Bridge.

A cursory examination of the map of Scotland will show that the estuaries of the Forth, Tay and Moray Firths have relatively short water crossings, but require considerable distances to be travelled by land. In the mid nineteenth century there was tremendous activity building railways. The first of the Scottish estuaries to be crossed was the Tay. This is over 1 km wide, but the water is comparatively shallow and a cast iron bridge was built and completed in 1878. A foundry was built on the south bank of the Tay to produce the castings.

In December 1879 the bridge fell down, killing nearly 300 passengers and crew, when a train fell from the high girders. The designer of the bridge was immediately accused of a poor design, incapable of withstanding the winds.

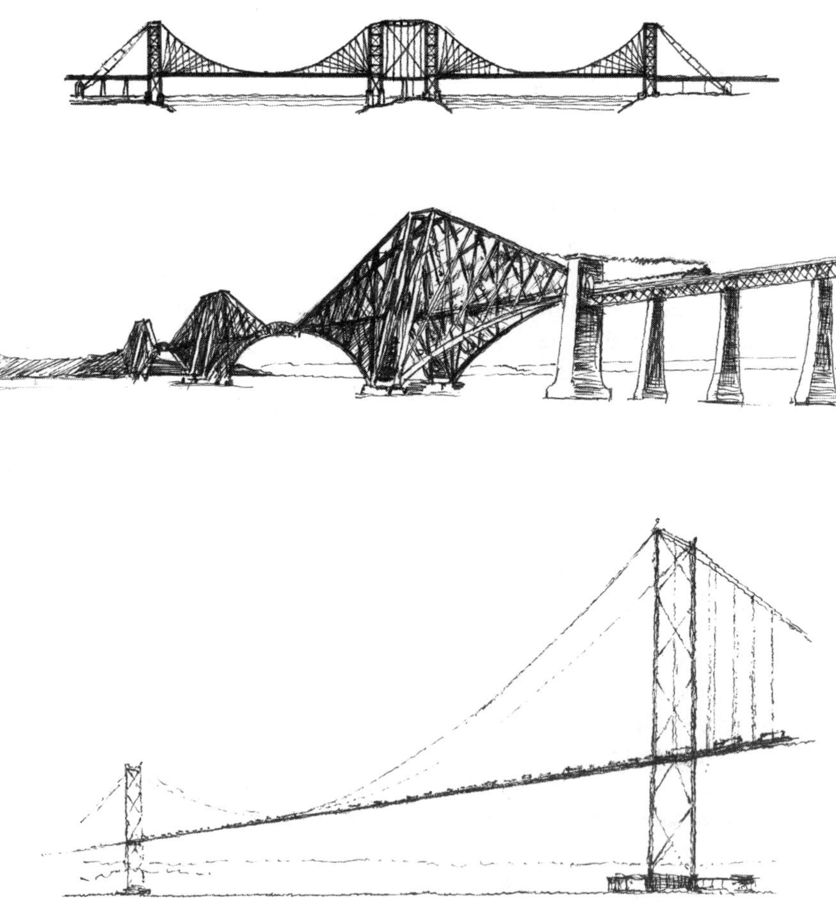

Figure 15.17 The design of the first Forth Railway Bridge, which was never built (top). The present Forth Railway Bridge (centre). The Forth Road Bridge (bottom).

The Forth estuary, although a shorter crossing, presented much more difficulty, with high banks on either side, much deeper water and the requirement of high clearance for shipping. For many years, designs had been produced either for tunnels or for bridges of various types.

When the Tay Bridge fell down, a railway bridge over the Forth was actually in the process of being built. It was designed and being supervised by the same man who built the Tay Bridge. Work was stopped immediately. This bridge would have used wrought iron, which was not available in quantity when the Tay Bridge was built, to produce an elegant design being a double suspension bridge.

Whether or not this bridge would have been successful will never now be known. What was produced instead was one of the wonders of the world – the Forth Railway Bridge – a massive cantilever design. This is the first major structure in the world to use mild steel and completed its centenary in 1991.

Parallel to this massive structure is the Forth Road Bridge of delicate proportions. Anyone looking at the minute train crossing the railway bridge and comparing this with the mass of traffic at busy times crossing the more slender road bridge must appreciate that either the road bridge should fall down or the safety factor of the railway bridge is rather excessive.

Going back to the case of the Tay Bridge, it is now known that the foundry produced low quality castings. A foundry foreman found that by using a paste obtained by mixing white-of-egg with cast iron filings, any porosity could be filled in to produce a visually satisfactory finish. This became known as 'Beaumont Egg', and it contributed little to the strength of the cast iron.

The riveting of the lattice beams was also inadequate. It is on record that a bridge inspector found missing rivets and instead of reporting them replaced them with rivets hand-tapped into position as he did not want to get anyone into trouble!

It is of some historical interest that, if a correct failure investigation had been carried out into the reason for the Tay Bridge failure, we might perhaps have had an elegant Forth Railway Bridge. On the other hand, this bridge might also have fallen down!

Index

Page numbers in *italics* refer to tables and those in **bold** refer to figures.

THE UNIVERSITY OF READING
LIBRARY